PLANT SECONDARY METABOLITES

Volume 3

Their Roles in Stress Ecophysiology

PLANT SECONDARY METABOLITES

Volume 3

Their Roles in Stress Ecophysiology

Edited by
Mohammed Wasim Siddiqui, PhD
Vasudha Bansal, PhD

Apple Academic Press Inc. | Apple Academic Press Inc.
3333 Mistwell Crescent | 9 Spinnaker Way
Oakville, ON L6L 0A2 | Waretown, NJ 08758
Canada | USA

©2017 by Apple Academic Press, Inc.

First issued in paperback 2021

Exclusive worldwide distribution by CRC Press, a member of Taylor & Francis Group
No claim to original U.S. Government works

ISBN 13: 978-1-77463-109-6 (pbk)
ISBN 13: 978-1-77188-356-6 (hbk)

Library and Archives Canada Cataloguing in Publication

Plant secondary metabolites.

Includes bibliographical references and indexes.
Contents: Volume 3. Their roles in stress ecophysiology / edited by Mohammed Wasim Siddiqui, PhD, Vasudha Bansal, PhD.
Issued in print and electronic formats.
ISBN 978-1-77188-356-6 (v. 3 : hardcover).--ISBN 978-1-77188-357-3 (v. 3 : pdf)
1. Plant metabolites. 2. Plants, Edible--Metabolism. 3. Medicinal plants--Metabolism.
4. Metabolism, Secondary. I. Siddiqui, Mohammed Wasim, author, editor II. Bansal, Vasudha, author, editor

QK881.P63 2016	572'.42	C2016-904969-8	C2016-904970-1

Library of Congress Cataloging-in-Publication Data

Names: Siddiqui, Mohammed Wasim, editor. | Prasad, Kamlesh, editor.
Title: Plant secondary metabolites / editors, Mohammed Wasim Siddiqui, Kamlesh Prasad.
Other titles: Plant secondary metabolites (Siddiqui)
Description: New Jersey : Apple Academic Press, Inc., [2017-] | Includes bibliographical references and index.
Identifiers: LCCN 2016030295 (print) | LCCN 2016031056 (ebook) | ISBN 9781771883528 (hardcover : alk. paper) | ISBN 9781771883535 (ebook) | ISBN 9781771883535 ()
Subjects: LCSH: Plant metabolites. | Plants, Edible--Metabolism. | Medicinal plants--Metabolism. | Metabolism, Secondary. | MESH: Plants, Edible--metabolism | Plant Extracts--chemistry | Phytochemicals | Plants, Medicinal Classification: LCC QK881 .P5526 2017 (print) | LCC QK881 (ebook) | NLM QK 98.5.A1 | DDC 572/.42--dc23
LC record available at https://lccn.loc.gov/2016030295

Apple Academic Press also publishes its books in a variety of electronic formats. Some content that appears in print may not be available in electronic format. For information about Apple Academic Press products, visit our website at **www.appleacademicpress.com** and the CRC Press website at **www.crcpress.com**

Plant Secondary Metabolites:

Volume 1: Biological and Therapeutic Significance

Editors: Mohammed Wasim Siddiqui, PhD, and Kamlesh Prasad, PhD

Plant Secondary Metabolites:

Volume 2: Stimulation, Extraction, and Utilization

Editors: Mohammed Wasim Siddiqui, PhD, Vasudha Bansal, PhD, and Kamlesh Prasad, PhD

Plant Secondary Metabolites:

Volume 3: Their Roles in Stress Ecophysiology

Editors: Mohammed Wasim Siddiqui, PhD, and Vasudha Bansal, PhD

ABOUT THE EDITORS

Mohammed Wasim Siddiqui, PhD

Dr. Mohammed Wasim Siddiqui is an Assistant Professor and Scientist in the Department of Food Science and Postharvest Technology, Bihar Agricultural University, Sabour, India, and the author or co-author of more than 33 peer-reviewed journal articles, 24 book chapters, and 18 conference papers. He has six edited and one authored books to his credit, published by Elsevier, USA; Springer, USA; CRC Press, USA; and Apple Academic Press, USA. Dr. Siddiqui has established an international peer-reviewed *Journal of Postharvest Technology.* He has been honored to be the Editor-in-Chief of two book series, Postharvest Biology and Technology and Innovations in Horticultural Science, both published by Apple Academic Press, New Jersey, USA, where he is a Senior Acquisitions Editor for Horticultural Science. He is also as an editorial board member of several journals.

Recently, Dr. Siddiqui received the Young Achiever Award 2014 for outstanding research work by the Society for Advancement of Human and Nature (SADHNA), Nauni, Himachal Pradesh, India, where he is also an Honorary Board Member. He has been an active member of organizing committees of several national and international seminars, conferences, and summits.

Dr. Siddiqui acquired BSc (Agriculture) degree from Jawaharlal Nehru Krishi Vishwa Vidyalaya, Jabalpur, India. He received MSc (Horticulture) and PhD (Horticulture) degrees from Bidhan Chandra Krishi Viswavidyalaya, Mohanpur, Nadia, India, with specialization in the Postharvest Technology. He was awarded a Maulana Azad National Fellowship Award from the University Grants Commission, New Delhi, India. He is a member of the Core Research Group at the Bihar Agricultural University (BAU), providing appropriate direction and assisting with sensiting priority of the research. He has received several grants from various funding agencies to carry out his research projects that are associated with postharvest technology and processing aspects of horticultural crops. Dr. Siddiqui is dynamically involved in teaching (graduate and doctorate students) and research, and he has proved himself as an active scientist in the area of Postharvest Technology.

Vasudha Bansal, PhD

Vasudha Bansal is working as a Postdoctoral Researcher in the Department of Environmental Engineering at Hanyang University in Seoul, South Korea. She has worked as a study coordinator in clinical trials for diabetic and osteoporotic patients in the Endocrinology Department in the Postgraduate Institute of Medical Education and Research (PGIMER) in Chandigarh, India. She has been awarded a young scientist talent scholarship from the Ministry of Education in Brazil (2014), a Bio-Nutra Junior award for best oral presentation from the National Institute of Food Technology and Entrepreneurship Management, Kundli, Haryana, India (2013), the Mrs. Gupta Physics award and the Mrs. Handa Zoology award for her BSc. She is an editorial member and peer reviewer of the *Journal of Food Research,* a peer reviewer of *Food Composition and Analysis* and the *Journal of Food Bioprocess and Technology*. She is a member of the Indian Science Congress Association, the Nutrition Society of India, and the Indian Dietetic Association. She acquired her PhD (food science) from the AcSIR-CSIO, Chandigarh (India) and her BSc and MSc in food and nutrition from Panjab University, Chandigarh, India.

DEDICATION

This Book
Is
Affectionately Dedicated
to Our Beloved Parents

CONTENTS

LIST OF CONTRIBUTORS

Mohammad Ansar
Plant Pathology, BAC, Bihar Agricultural University, Sabour, Bhagalpur 813210, Bihar, India.

Rekha Balodi
Plant Pathology, COA, Govind Ballabh Pant University of Agriculture and Technology, Udham Singh Nagar, Pantnagar 263145, Uttarakhand, India.

Vasudha Bansal
Department of Civil & Environmental Engineering, Hanyang University, 222 Wangsimni-Ro, Seoul 133791, South Korea.

Satya Sundar Bhattacharya
Department of Environmental Science, Tezpur University, Tezpur 784028, Assam, India. E-mail: satyasundarb@yahoo.co.in; satya72@tezu.ernet.in.

Nithya C.
Division of Entomology, Indian Agricultural Research Institute (IARI), New Delhi 110012, India.

Subhasish Das
Department of Environmental Science, Tezpur University, Tezpur 784028, Assam, India.

Abhijeet Ghatak
Plant Pathology, BAC, Bihar Agricultural University, Sabour, Bhagalpur 813210, Bihar, India.

Lajja Vati Ghatak
Plant Breeding and Genetics, BAC, Bihar Agricultural University, Sabour, Bhagalpur 813210, Bihar, India. E-mail: ghatak11@gmail.com.

Norsuhada Abdul Karim
Bioprocess Engineering Department, Faculty of Chemical Engineering, Universiti Teknologi Malaysia, 81310 Johor Bahru, Johor, Malaysia.

Pawan Kumar
Department of Chemical Engineering, Indian Institute of Technology, Hauz Khas, New Delhi 110016, India.

Kiran Kumari
Department of Entomology, Bihar Agricultural University, Sabour, Bhagalpur 813210, Bihar, India.

Kalmesh M.
Department of Entomology, Bihar Agricultural University, Sabour, Bhagalpur 813210, Bihar, India.

Ida Idayu Muhamad
Bioprocess Engineering Department, Faculty of Chemical Engineering, Universiti Teknologi Malaysia, 81310 Johor Bahru, Johor, Malaysia; IJN-UTM Cardio Engineering Centre, V01 FBME, Universiti Teknologi Malaysia, 81310 Johor Bahru, Johor, Malaysia.

Kamlesh Prasad
Department of Food Engineering and Technology, Sant Longowal Institute of Engineering & Technology, Longowal 148106, Punjab, India.

Chandramani Raj
Dryland Cereal Pathology, International Crop Research Institute for Semi-Arid Tropics, Patancheru, Hyderabad 502324, Telangana, India.

S. N. Ray
Department of Entomology, Bihar Agricultural University, Sabour, Bhagalpur 813210, Bihar, India.

Shyambabu S.
Department of Entomology, Bihar Agricultural University, Sabour, Bhagalpur 813210, Bihar, India.

Farzaneh Sabbagh
Bioprocess Engineering Department, Faculty of Chemical Engineering, Universiti Teknologi Malaysia, 81310 Johor Bahru, Johor, Malaysia.

Tamoghna Saha
Department of Entomology, Bihar Agricultural University, Sabour, Bhagalpur 813210, Bihar, India. E-mail: tamoghnasaha1984@gmail.com.

Mohammed Wasim Siddiqui
Department of Food Science and Postharvest Technology, Bihar Agricultural University, Sabour, Bhagalpur 813210, Bihar, India.

Ashwathnarayan Srinivasaraghavan
Plant Pathology, BAC, Bihar Agricultural University, Sabour, Bhagalpur 813210, Bihar, India.

Satish K. Tuteja
Academy of Scientific and Innovative Research, CSIR-Central Scientific Instruments Organisation, Sector-30, Chandigarh 160030, India.

LIST OF ABBREVIATIONS

3D	three-dimensional
3HA	3-hydroxyalkanoates
3HB	3-hydroxybutiric acid
AAs	amino acids
AEC	anion exchange capacity
AOX	alternative oxidase
BRs	brassinosteroids
CEC	cation exchange capacity
CNS	central nervous system
EU	European Union
FT-IR	Fourier transform-infrared spectroscopy analysis
GA3	gibberellin
GC	gas chromatography
HCN	hydrogen cyanide
HPLC	high performance liquid chromatography
HSPs	heat shock proteins
IAA	indole-3-acetic acid
IPM	integrated pest management
lcl-PHAs	long chain-length PHAs
NMR	nuclear magnetic resonance
OM	organic matter
PHAs	polyhydroxyalkanoates
POXs	peroxidases
ROS	reactive oxygen species
SAR	structure–activity relationship
scl-PHAs	short-chain-length PHAs
TEM	transmission electron microscopy
TGs	triglycerides
VOCs	volatile organic compounds

PREFACE

Secondary metabolites are known to protect plants from adverse environmental conditions. The stress conditions stimulate the production or liberation of secondary metabolites. These metabolites do not render the growth factor for plants but rather ensure their defensive mechanism. The adverse conditions of adaptability cause the release of bioactive compounds. The relationship between plant physiology and the release of their accumulated secondary metabolites under stressed conditions has been shown in a comprehensive way.

This volume, *Plant Secondary Metabolites: Their Roles in Stress Ecophysiology,* has been contributed to by experienced experts in the respective fields. Chapter 1 describes the relationship between environmental stress and physiology of plants, leading to stimulation of secondary metabolites; Chapter 2 covers the significance and effect of organic matter present in soil toward the secondary metabolites accumulated in plants; Chapter 3 illustrates the interactions between soil and plants under the different zones of climate; Chapter 4 discusses the essential nutritional aspects present in plants and the effect of soil composition on the secondary metabolites; Chapter 5 elaborates on the application of secondary metabolites in pest management; Chapter 6 exhibits the effect of pathogens on the release of secondary metabolites; Chapter 7 presents the secondary metabolite as the innovative product 'polyhydroxyalkanoates' produced as a by-product by the microorganisms; and Chapter 8 precisely discusses the diverse utilization of secondary metabolites from plant sources.

The book is intended to draw attention toward the adverse climatic factors and wilted plants, which can be maneuvered for the application of the secondary metabolites to various industrial fields. The book will prove a useful venture for researchers working in stress physiology, soil–plant relationships, and natural secondary metabolites.

The editors welcome suggestions and remarks from readers to develop/ update the future edition.

ACKNOWLEDGMENTS

It was almost impossible to express the deepest sense of veneration to all without whose precious exhortation this book project could not be completed. At the onset of the acknowledgment, we ascribe all glory to the gracious "Almighty God" from whom all blessings come. We would like to thank for His blessing to write this book.

With a profound and unfading sense of gratitude, we sincerely thank the Bihar Agricultural University (BAU), India, CSIR—Central Scientific Instruments Organisation (CSIO), India, and Sant Longowal Institute of Engineering & Technology (SLIET), India, for providing us the opportunity and facilities to execute such an exciting project, and for supporting us toward our research and other intellectual activities around the globe. We convey special thanks to our colleagues and other research team members for their support and encouragement for helping us in every footstep to accomplish this venture. We would like to thank Mr. Ashish Kumar, Ms. Sandy Jones Sickels, and Mr. Rakesh Kumar of Apple Academic Press for their continuous support to complete the project.

Our vocabulary will remain insufficient in expressing our indebtedness to our beloved parents and family members for their infinitive love, cordial affection, incessant inspiration, and silent prayer to "God" for our well-being and confidence.

CHAPTER 1

ENVIRONMENTAL STRESS AND STRESS BIOLOGY IN PLANTS

SUBHASISH DAS and SATYA SUNDAR BHATTACHARYA*

Department of Environmental Science, Tezpur University, Tezpur 784028, Assam, India

*Corresponding author, Tel.: +91 3712 267007/+91 3712 2670078/+91 3712 2670079x5610; E-mail: satyasundarb@yahoo. co.in; satya72@tezu.ernet.in.

CONTENTS

ABSTRACT

The various external and internal stress factors in the form of abiotic and biotic stress lead to the changes in the physiological metabolism of the plants. This puts an effect on the mechanism of the inbuilt features (in terms of growth, reproductive patterns, and genetic patterns). Therefore, the number of factors such as effect of temperature, heat, and photosynthesis on the seedling development, hormone secretion, and plant growth were showcased.

1.1   AN OVERVIEW OF PLANT STRESS FACTORS

Terrestrial plants are sessile in nature and, thus, are vulnerable to any abrupt environmental changes. These changes are of varied nature and form. For instance, changes in soil or aerial environment constitute the abiotic stress source while pest attack, microbial infestation, overgrazing, anthropogenic influences, and so on, form the types of biotic stresses. In order to tackle with most of the stress situations, certain mechanisms in plants are developed by altering or modifying their physiology, growth, and reproductive patterns. As such, various molecular mechanisms are developed by plants in order to overcome stress.

Plants which are better adapted to their environment survive the stress conditions while the others simply perish. Genes have a significant role to play in regulating the physiology of plants as they act as the precursor for various enzyme and metabolite synthesis. These metabolites act as stress busters which soothe the plant of its stresses and finally homeostasis is created between the plant and its environment. Speaking of environment, soil and climate form the major domains of stress elicitors in plants. The soil attributes like water, nutrients, organic matter, and so on, along with its very physical makeup are at the forefront for stress enhancers in plants. Various plants simply die due to drought or nutrient stress while some die because of high soil compaction or lower organic matter availability which directly affects their phenology. The climatic stress factors include solar insolation intensity, temperature fluctuation, precipitation pattern, wind velocity, and so on, which directly influence plant physiology and thereby their very existence. Over a long time, plants have developed different strategies like nutrient and water storage, changes in morphology, alteration in flowering time and patterns, root architecture, and so on, to mitigate stress conditions. In the following sections, various types of environmental stresses, especially soil-mediated plant stresses have been discussed along with plant stress biology.

1.2 TEMPERATURE STRESS AND PLANT RESPONSE

Plants encounter massive fluctuations in temperature that poses enormous stress on their normal physiology. Extreme temperature conditions like heat, cold, or frost directly affect the plant cellular structures leading to their disruption or and damage of cellular organelles (Rurek, 2014). Various plant proteins get denatured and inactive leading to high oxidative stresses due to temperature extremes. Workers throughout the world have reported significant decline in crop productivity due to ambient temperature fluctuations (Nahar et al., 2009; Bita & Gerats, 2013; Ntatsi et al., 2014).

Plant temperature is mostly dependent on the ambient air temperature and solar influx. Different plants have different optimum temperature range for normal functioning which varies with respect to the environmental conditions. As such, a temperature range harmful for a plant in a particular place may be essential to some in other places (Krasensky & Jonak, 2012). According to Levitt (1980), plants implement two major strategies to combat temperature stress: stress avoidance and stress tolerance. Stress avoidance deals with mechanisms which resist the harmful impact of any environmental factor on plant. For example, epiphytes which have modified their morphology and physiology to adapt to the environment, and these developments are inherited to the upcoming generations. On the other hand, stress tolerance involves certain processes by which plants tune themselves to the environmental changes by modifying their physiology during the stress period. For example, production of some membrane protective proteins by plants which check cellular disruption during heat stress. Unlike stress avoidance mechanism, stress tolerance is a reversible process. The probable impact of various environmental stresses on plant physiology is presented in Figure 1.1.

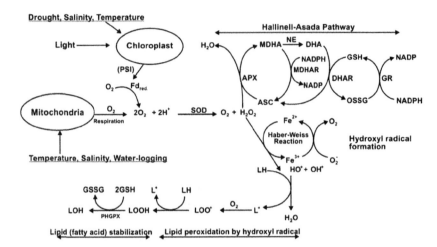

FIGURE 1.1 Schematic diagram illustrating the impact of various environmental stresses on generation and scavenging of superoxide radical, hydrogen peroxide, hydroxyl radical-induced lipid peroxidation, and glutathione POX-mediated fatty acid stabilization under environmental stresses. APX, ascorbate POX; ASC, ascorbate; DHA, dehydroascorbate; DHAR, dehydroascorbate reductase; Fd, ferredoxin; GR, glutathione reductase; GSH, red glutathione; GSSG, oxi-glutathione; HO, hydroxyl radical; LH, lipid; L, LOO; LOOH, unstable lipid radicals and hydroperoxides; LOH, stable lipid (fatty acid); MDHA, monodehydro-ascorbate; MDHAR, mono dehydro-ascorbate reductase; NE, non-enzymatic reaction; PHGPX, phospholipid-hydroperoxide glutathione POX; SOD, superoxide dismutase. (From Wahid, A., Gelani, S., Ashraf, M., Foolad, M. R., Environ. Exp. Bot., 61, 208, 2007. Reprinted with permission from Elsevier).

1.2.1   HEAT SHOCK PROTEINS

Proteins are a vital component of the plant physiological system. Under normal conditions, proteins function efficiently. However, under conditions of high temperature, proteins tend to fold and degenerate. Heat stress as well as other stresses can initiate certain defensive mechanisms like expression of particular genes that were not expressed under "normal" conditions (Morimoto, 1993; Feder, 2006). The response to stresses on the molecular level is prevalent in all living things, especially those changes in genotypic expression giving rise to an increase in the synthesis of protein groups. These groups are termed popularly as "heat shock proteins" (HSPs), "stress-induced proteins," or "stress proteins" (Lindquist & Crig, 1988; Morimoto et al., 1994; Gupta et al., 2010). The HSPs differ in their molecular weights and are expressed in times when the plant experiences high temperature. The HSPs perform various functions in the plant body. For example some HSPs are constitutive while some are inducible in function. HSP100 is the most commonly expressed protein in plant body under high temperature stress (Queitsch et al., 2000). Subsequent research has revealed that most HSPs have strong cytoprotective effects and act as molecular chaperones for other cellular proteins. Inappropriate activation of signaling pathways could occur during acute or chronic stress as a result of protein misfolding, protein aggregation, or disruption of regulatory complexes. The major action of chaperones is thought to restore balance in protein homeostasis. Mammalian HSPs can be differentiated into five families, according to their molecular size: HSP100, HSP90, HSP70, HSP60, and the small HSPs. Each family of HSPs is composed of members expressed either constitutively or regulated inductively and are targeted to different subcellular compartments. For example, although HSP90 is constitutively abundantly expressed in the cells, HSP70 and HSP27 are highly induced by different stresses such as heat, oxidative stress, or anticancer drugs (Schmitt et al., 2007). According to Levitt et al. (1997), the function of any protein is ascertained by its formation and folding into 3-dimensional structures. The development of three dimensional structures requires 50% of primary amino acids (AAs) sequence (Dobson et al., 1998). Thus, HSPs are significant in the folding of other proteins. Reports highlight that HSPs protect cells from injury and facilitate recovery and survival after a return to normal growth conditions (Morimoto & Santoro, 1998). On the contrary, Timperio et al. (2008) showed that upon heat stress, the role of HSPs as molecular chaperones is undoubtedly prominent whereas their functions could be different in nonthermal stress conditions. Unfolding of proteins is not the main effect and protection from damage could occur in an alternative way apart from ensuring the maintenance of correct protein structure (Al-Whaibi, 2011). Figure 1.2 presents a schematic diagram illustrating the effect of heat stress on plant physiology.

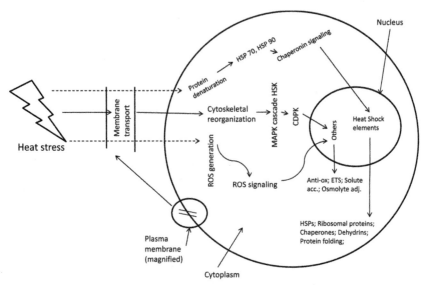

FIGURE 1.2   Schematic diagram showing the heat stress response. MAPK, mitogen activated protein kinases; ROS, reactive oxygen species; HAMK, heat shock activated MAPK; HSE, heat shock element; HSPs, heat shock proteins; CDPK, calcium-dependent protein kinase; HSK, histidine kinase.

1.2.2   INFLUENCE OF HIGH-TEMPERATURE STRESS ON PLANTS

Heat stress can affect some plant processes more than others. Extreme temperature influences various processes, but the most important effects are those that are first encountered as temperatures rise above the optimum for plant growth. Two plant processes that are particularly sensitive to heat stress are pollen development and photosynthesis. Other processes that appear to be inherently less sensitive to heat stress include respiration (Berry & Raison, 1981). Pollen development and fruit set are important as they give rise to the harvestable parts of plant. The effect of heat stress on crop yield will depend upon the timing of the heat stress. If the stress is experienced during anthesis, substantial loss in fruit set and, ultimately, crop yield can occur. Photosynthesis, on the other hand, is also particularly sensitive to heat stress and heat stress at nearly any time during crop growth could have adverse consequences on yield through stress effects on photosynthesis (Guilioni et al., 2003). Because pollen development and photosynthesis are the two most high-temperature-sensitive plant processes, the next sections will cover the effects of high temperature on these processes.

1.2.1.1   EFFECT ON POLLEN DEVELOPMENT

Various workers have mentioned the influence of heat stress on processes of pollen development in plants. Prasad et al. (2002) subjected common bean (*Phaseolus vulgaris* L.) to five temperature regimes between 28°C/18°C day/night and 40°C/30°C day/night. These plants exhibited substantial loss in pod set above 37°C/27°C day/night temperature and reductions in seed set at even lower temperature. Pollen viability was even more sensitive and was more than 50% inhibited in the plants grown at 37°C/27°C day/night temperature. These effects were not sensitive to increased CO_2, unlike some effects of temperature on photosynthesis (Cowling & Sage, 1998). This is consistent with the effects being primarily on pollen viability and development of reproductive structures rather than on photosynthesis.

1.2.1.2   EFFECT ON SEEDLING ESTABLISHMENT

Heat can limit plant growth at the seedling stage, because the temperature near the soil can be very high as a result of a boundary layer of air near the soil surface (Campbell & Norman, 1998). Soil temperature can surpass 50°C when the sun is high and bright for a longer period. High soil temperature hinders in root development and leaf setting in tender seedlings. Metabolism in seedlings of many species can respond to heat through the induction of HSPs (Vierling, 1991). Many of the studies of HSP-derived thermotolerance use assays of seedling establishment (Hong & Vierling, 2000; Queitsch et al., 2000; Hong et al., 2003), which is appropriate, even though HSPs have effects well beyond their effect on this phase of the plant's development.

1.2.1.3   EFFECT ON PHOTOSYNTHESIS

Primary metabolic events of plants like photosynthesis is under direct influence of ambient temperature. Even in the absence of any injury, photosynthesis of C3 plants would be expected to decline as temperature increases because photorespiration increases comparatively faster than photosynthesis with elevating temperature (Schuster & Monson, 1990). Various reports reveal that heat directly damages the photosynthetic apparatus, with photosystem II (PSII) often considered a key weak link (Santarius, 1975; Santarius & Müller, 1979; Berry & Björkman, 1980), but specifically above 45°C (Terzaghi et al., 1989; Thompson et al., 1989; Gombos et al., 1994; Cajánek et al., 1998). One of the major incidences of high temperature is destruction of the oxygen-evolving complex with the loss of a 33 kDa extrinsic protein and Mn^{2+} (Enami et al., 1994). However, at temperatures <40°C some reversible reductions in PSII-dependent electron transport can be seen while irreversible effects and loss of the 33 kDa proteins occur at temperatures >42°C (Yamane et al., 1998). Thus, damage to PSII cannot explain widely observed, heat-induced depression in photosynthesis seen at temperatures between 35 and 40°C.

1.2.3 INFLUENCE OF LOW TEMPERATURE ON PLANTS

In nature, cold acclimation is initiated by a combination of the shortening of the photoperiod which results in growth cessation and temperatures less than 10°C. Similar to high temperature stress, low temperatures also render plants huge morphological deformities coupled with prominent physiological malfunctioning. Some species are more sensitive to low temperature induction than others. For example in a cold acclimation experiment winter rye (*Secale cereale*) increased in freezing tolerance when exposed to 10°C, whereas winter wheat (*Triticum aestivium*) required 7°C to initiate cold acclimation. In contrast, spring wheat did not increase in freezing tolerance until exposed to 4°C. Thus, there appears to be differences in response to low-temperature signaling. Exposure to other abiotic stresses (such as drought, UVB, wind, etc.) in the field at warm temperatures can also increase freezing. Under natural conditions plants germinate, grow, and mature under a constant state of environmental fluxes. Daily temperature variation can be as much as 10–15°C; however, it is very rare that a large temperature change of 20–2°C occurs over a matter of a few minutes, especially in the rhizosphere. Soil has a buffering capacity to normalize the conditions which efficiently manages the changes in soil temperature and water availability resulting in a stable state. The following is a list of shock-associated proteins; chitinases, peroxidases (POXs), and PR-1 like protein (Mackerness et al., 2001); cytochrome P450, a C_2H_2-type zinc-finger protein and a blue copper-binding protein (Desikan et al., 2001) and proteins associated with reactive oxygen species (ROS) (Gonzalez-Meler et al., 2001).

1.3 DROUGHT STRESS AND PLANT RESPONSE

Drought stress is one of the most adverse factors of plant growth and productivity and regarded as a severe threat for sustainable crop production in the scenario of changing climate. Drought triggers a wide variety of plant responses, ranging from cellular metabolism to changes in growth rates and crop yields. Understanding the biochemical and molecular responses to drought is essential for a holistic perception of plant resistance mechanisms to water-limited conditions. Drought, being the most important environmental stress, severely impairs plant growth and development, limits plant production, and the performance of crop plants more than any other environmental factor (Shao et al., 2009). Plant experiences drought stress either when the water supply to roots becomes difficult or when the transpiration rate becomes very high. Available volume of utilizable water for successful crop production has fallen in recent years. Furthermore, under the context of various climatic change models, scientists suggested that in many regions of world, crop losses due to increasing water shortage will further worsen its impacts. Some of the major impacts of drought conditions leading to water stress can be seen in the following subsections.

1.3.1 IMPACT OF DROUGHT ON PLANT PHYSIOLOGY

Plants cope with drought stress by manipulating key physiological processes like photosynthesis, respiration, water relations, antioxidant, and hormonal metabolism. Some of the major impacts are discussed as under.

1.3.1.1 RELATION BETWEEN PLANT GROWTH AND WATER AVAILABILITY

A primary response of plants subjected to drought stress is growth arrest. During drought conditions, the shoot growth is inhibited which in turn reduces metabolic demands of the plant and mobilizes metabolites for the production of protective compounds required for osmotic balance. Root-growth arrest enables the root meristem to remain functional and gives rise to rapid root growth when the stress is relieved (Hsiao & Xu, 2000). Lateral root inhibition has also been seen to be an adaptive response, which results in growth promotion of the primary root, enabling extraction of water from the lower layers of soil (Xiong et al., 2006). Occurrence of growth inhibition starts due to the loss of cell turgor arising from the lack of water availability to the growing cells. The cellular water availability decreases due to poor hydraulic conductance from roots to leaves caused by stomatal closure. Although a decrease in hydraulic conductance decreases the supply of nutrients to the shoot, it also prevents embolism in xylem and could form an adaptive response to mitigate drought stress. Adjustment of the osmotic balance is another way by which plants cope with drought stress. Synthesis of compatible solutes like polyols and proline under stress prevents the water loss from cells and plays an important role in turgor maintenance (Blum, 2005, DaCosta & Huang, 2006). Modification of growth processes coupled with reduction in the performance of photosynthetic organs under stress conditions leads to subsequent changes in carbon partitioning between the source and sink tissues (Roitsch, 1999). Hence, carbohydrates that contribute to growth under normal growth conditions become easily available for selective growth of roots or for the synthesis of solutes for osmotic adjustment (Lei et al., 2006, Xue et al., 2008).

1.3.1.2 *RELATION BETWEEN PHOTOSYNTHESIS AND WATER AVAILABILITY*

The physiological pathways of photosynthesis in plants depend on water significantly. Water deficit-induced ABA synthesis brings about stomatal closure, which results in a decrease in intercellular carbon dioxide concentration and inhibits photosynthesis. This inhibition is temporary and can be resumed if stomata open upon stress removal (Chaves et al., 2009). On the contrary, open stomata and high-hydraulic conductance under drought permit photosynthesis and nutrient supply to the shoot by risking turgor loss (Sade et al., 2012). Some plants tend to follow the latter strategy to peruse the synthesis of osmotic metabolites from photoassimilates, which aid in preventing turgor loss. Carbon dioxide limitation due to prolonged stomatal closure in the visage of continued photosynthetic light reactions leads to the accumulation of reduced photosynthetic electron transport components, which can reduce molecular oxygen and ultimately give rise to ROS, resulting in indiscriminate damage to the photosynthetic apparatus. This metabolic inhibition of photosynthesis is irreversible and leads to injury (Lawlor & Cornic, 2002). Due to such temporary inhibition, photo-phosphorylation and ATP generation is reduced, which restricts Ribulose-1,5-bisphosphate carboxylase/oxygenase (RuBisCO) activity. The major adaptive responses to prevent drought-induced damage to photosynthetic apparatus involve thermal dissipation of light energy, photodestruction of D1 protein of PSII, the xanthophyll cycle, water–water cycle, and disso-ciation of the light-harvesting complexes from photosynthetic reaction centers (Niyogi, 1999, Demmig-Adams and Adams, 2006).

1.3.1.3 RELATION BETWEEN PLANT RESPIRATION AND WATER AVAILABILITY

Plant respiration is an invincible part of plant physiology that modifies its growth pattern. The ratio between photosynthetic CO_2 assimilation and respiratory CO_2 release is determinant of plant growth in a given condi-tion. The rate of respiration is regulated by processes that use the respira-tory products—ATP (water and solute uptake by roots, translocation of assimilates to sink tissues), NADH, and tricarboxylic acid (TCA) cycle intermediates (biosynthetic processes in growing parts of a plant), which together contribute to plant growth. Under drought stress, these processes are affected and result in diminishing rate of respiration. Consequently, incidences of increased respiratory rates have also been observed under water scarcity, and these give rise to an increase in the intercellular CO_2 levels in leaves (Lawlor & Tezara, 2009). Elevated respiration may arise due to uncoupling of respiratory oxygen evolution from oxidative phos-phorylation, which resists the accumulation of reductants and reduces the evolution of ROS. Increased respiratory rates are also observed due to the activation of energy-intensive processes like osmolyte synthesis and antioxidant metabolism that occur under drought conditions. Various reports highlight the interdependence of metabolic processes in chloro-plasts and mitochondria (Raghavendra & Padmasree, 2003). For example, mitochondria are involved in processing the glycolate produced in chloroplasts during photorespiration (Taira et al., 2004). Mitochondrial respiration also plays an important role in dissipating the NADPH molecules formed during photosynthetic light reactions through type II NADPH dehydrogenases located on the matrix side (Plaxton & Podesta, 2006). In such situations, leaf mitochondria act as a safety engine that enables the plant to cope with fluctuations in chloroplast metabolism under water stress (Atkin & Macherel, 2009). Plant mitochondria also resist the generation of ROS within themselves by employing the alternative oxidase (AOX) pathway, wherein the complexes III and IV of the respiratory electron transport system are bypassed and electrons are directly transferred to oxygen, with the liberation of thermal energy instead of ATP (Siedow & Umbach, 2000). The AOX pathway as well as the photorespiratory pathway becomes functional when a plant is exposed to stress and serves a role in maintaining cell function by preventing the accumulation of ROS (Lambers et al., 2005; Florez-Sarasa et al., 2007). In addition, the TCA cycle is modified to prevent the generation of excess reductants. One of the modifications is GABA synthesis, in which two steps in the TCA cycle related to the generation of reducing power are bypassed. The accumulation of GABA occurs during stress conditions and may involve a stress adaptive response (Fait et al., 2007). Large protein complexes called "prohibitins" found in the inner mitochondrial membrane play a role in maintaining the superstructure of the inner mitochondrial membrane and the protein complexes associated with it (Van Aken et al., 2010). They have been implied in stress tolerance not only because of their role in protecting mitochondrial structure, but also in triggering retrograde signaling between mitochondria and the nucleus in response to stress, thus altering the expression of several stress-responsive transcripts, including AOX, HSPs, and genes involved in hormone homoeostasis.

1.3.1.4 RELATIONSHIP BETWEEN PLANT ANTIOXIDANTS AND WATER AVAILABILITY

ROS are generated due to metabolic perturbation of cells, and these cause cell damage and death. While mechanisms to prevent the generation of ROS have been mentioned earlier, an important adaptive mechanism involves their effective scavenging if and when these harmful species do arise. Antioxidant substrates like ascorbate, α-tocopherol, and carotenoids, and antioxidant enzymes like superoxide dismutase, catalase, ascorbate POX, and glutathione reductase exist in cell organelles and the cytoplasm and play a pivotal role in detoxifying these reactive species (Shao et al., 2008). Another class of antioxidant enzymes, methionine sulfoxide reductases, plays a role in preventing damage to proteins due to ROS generation in plastids (Rouhier et al., 2006). These enzymes use thioredoxin to reduce the methionine sulfoxide residues generated in proteins due to oxidative stress.

1.3.1.5   RELATIONSHIP BETWEEN PLANT HORMONES AND WATER AVAILABILITY

Plant hormones regulate diverse processes in plants, which enable acclimation to stress. When plants experience water deficits, ABA is synthesized in roots and translocated to leaves, where it brings about stomatal closure and inhibits plant growth, thus enabling the plant to adapt to stress conditions (Wilkinson & Davies, 2010). Thameur et al. (2011) recorded a fivefold increase in endogenous ABA levels in drought-tolerant varieties of barley as compared to susceptible ones, indicating its role in improving stress tolerance. The role of ABA in regulating aquaporin activity, which contributes to the maintenance of a favorable plant water status, has also been reported (Parent et al., 2009). Improvement of shoot growth under drought was observed when 9-cis-epoxycarotenoid dioxygenase (NCED3), a key enzyme in abscisic acid biosynthesis, was overexpressed in *Arabidopsis* (Iuchi et al., 2001). Thus drought events lead to ABA accumulation that further brings about a reduction in ethylene production and an inhibition of ethylene-induced senescence and abscission. Sharp (2002) reported an increase in ethylene production and low susceptibility to drought conditions with ABA-deficient maize seedlings. Auxins have been identified as negative regulators of drought tolerance. In wheat leaves, drought stress tolerance was accompanied by a decrease in indole-3-acetic acid (IAA) content (Xie et al., 2003). Down regulation of IAA was seen to facilitate the accumulation of late embryogenesis-abundant mRNA, leading to drought stress adaptation in rice (Zhang et al., 2009). However, there are evidences of a transient increase in IAA content in maize leaves during the initial stages of exposure to water stress, which later drops sharply as the plant acclimates to water stress. A rapid decline in endogenous hormones like zeatin and gibberellin (GA3) was also observed in maize leaves subjected to water stress, which correlated with higher levels of cell damage and plant growth inhibition. Reduced cytokinin content and activity caused by either reduced biosynthesis or enhanced degradation was observed in drought-stressed plants (Pospisilova et al., 2000). In alfalfa, decreased cytokinin content during drought led to accelerated senescence (Goicoechea et al., 1995). Cytokinins are known to delay senescence, and an increase in the endogenous levels of cytokinins through the overexpression of the *ipt* gene involved in cytokinin biosynthesis led to stress adaptation by delaying drought-induced senescence (Peleg & Blumwald, 2011). Cytokinins are also negative regulators of root growth and branching, and root-specific degradation of cytokinin contributed to primary root growth and branching induced by drought stress, hence increasing drought tolerance in *Arabidopsis* (Werner et al., 2010). Brassinosteroids (BRs) have also been reported to protect plants against various abiotic stresses (Kagale et al., 2007). Application of BR was seen to increase water uptake and membrane stability, as well as to reduce ion leakage arising from membrane damage in wheat plants subjected to drought stress (Sairam, 1994). However, it was shown that changes in endogenous BR levels did not occur during the exposure of pea plants to water stress (Jager et al., 2008).

1.4   SALINITY STRESS AND PLANT RESPONSE

Highly saline soils (i.e., soil with excessive salt concentration) affect plant physiology in a drastic way. Salinity stress involves changes in various physiological and metabolic processes, depending on severity and duration of the stress, and ultimately inhibits crop production (James et al., 2011; Rahnama et al., 2010; Munns, 2005; Rozema & Flowers, 2008). Initially, soil salinity resists plant growth in the form of osmotic stress which is then followed by ion toxicity (James et al., 2011; Rahnama et al., 2010). Initial phases of salinity stress involve decreased water absorption capacity of root systems and subsequent increase in water loss from leaves due to osmotic stress of high salt accumulation in soil and plants; and therefore, salinity stress is also considered as hyperosmotic stress (Munns, 2005). Osmotic stress in the initial stage of salinity stress causes various physiological changes such as interruption of membranes, nutrient imbalance, impairs the ability to detoxify ROS, brings about differences in the antioxidant enzymes and decreased photosynthetic activity, and decrease in stomatal aperture (Munns & Tester, 2008; Rahnama et al., 2010). Salinity stress is also considered as a hyperionic stress. One of the most detrimental effects of salinity stress is the accumulation of Na^+ and Cl^- ions in tissues of plants exposed to soils with high NaCl concentrations. Significant rise in production of ROS in plants (singlet oxygen, superoxide, hydroxyl radical, and hydrogen peroxide) has been reported due to salinity stress by many workers (Apel & Hirt, 2004; Mahajan & Tuteja, 2005; Ahmad, 2010; Ahmad & Prasad, 2012; Ahmad & Umar, 2011). Salinity-induced ROS formation can lead to oxidative damages in various cellular components such as proteins, lipids, and DNA, interrupting vital cellular functions of plants.

1.4.1 IMPACT OF SALT STRESS ON PLANT PHYSIOLOGY

Tolerance against salt stress needs profound changes in gene expression which is accompanied with changes in composition of plant transcriptome, metabolome, and proteome. Changes in gene expression at transcript level cannot exactly show the changes at protein level. This reflects the high importance of plant proteome since proteins are directly involved in plant stress response. In addition to enzymes, proteins include components of transcription and translation machinery. Therefore, they can regulate plant stress response at transcript and protein levels (Kosova et al., 2011).

1.4.1.1 INFLUENCE OF SALINITY STRESS ON PLANT CELLULAR MECHANISMS

High salinity causes hyperosmotic stress and ion disequilibrium that produce secondary effects or pathologies (Hasegawa et al., 2000; Zhu, 2001). Fundamentally, plants cope by either avoiding or tolerating salt stress. Plants remain either dormant during the salt episode or they tolerate salt stress through cellular adjustments. Tolerance mechanisms can be categorized as those that function to (1) minimize osmotic stress, (2) reduce the ion disequilibrium, or (3) alleviate the consequent secondary effects caused by these stresses. The chemical potential of the saline solution initially establishes a water potential imbalance between the apoplast and symplast that leads to turgor decrease, which if severe enough, can cause growth reduction (Bohnert et al., 1995). Growth arrest occurs when turgor is reduced below the yield threshold of the cell wall. Cellular dehydration begins when the water potential difference is greater than that can be compensated for by turgor loss (Taiz & Zeiger, 1998).

1.4.1.2 ROLE OF OSMOLYTES AND OSMOPROTECTANTS DURING SALINITY STRESS

As indicated previously, salt tolerance requires some compatible solutes to be accumulated in the cytosol and organelles where the solutes function in osmotic adjustment and osmoprotection (Rhodes & Hanson, 1993). Some compatible osmolytes are essential elemental ions, such as K^+, with the majority of organic solutes. Compatible solute accumulation as a response to osmotic stress is a ubiquitous process in organisms as diverse as bacteria to plants and animals. The accumulation of solutes varies with the organism and even between plant species. A major category of organic osmotic solutes consists of simple sugars (mainly fructose and glucose), sugar alcohols (glycerol and methylated inositols), and complex sugars (trehalose, raffinose, and fructans) (Bohnert & Jensen, 1996). Other solutes include quaternary AA derivatives (proline, glycine betaine, β-alanine betaine, proline betaine), tertiary amines (1,4,5,6-tetrahydro-2-mehyl-4-carboxyl pyrimidine), and sulfonium compounds (choline osulfate, dimethyl sulfonium propironate) (Nuccio et al., 1999). Many organic osmolytes are presumed to be osmoprotectants, as their levels of accumulation are insufficient to facilitate osmotic adjustment. Glycine betaine preserves thylakoid and plasma membrane integrity after exposure to saline solutions or to freezing or high temperatures (Rhodes & Hanson, 1993). Furthermore, various osmoprotectants elevate stress tolerance of plants when expressed as transgene products (Bohnert & Jensen, 1996; Zhu, 2001). One of the biochemical functions of osmoprotectants is the scavenging of ROS that are by-products of hyperosmotic and ionic stresses and cause membrane dysfunction and cell death (Bohnert & Jensen, 1996).

1.4.1.3 IMPORTANCE OF ION HOMEOSTASIS IN MITIGATION OF SALINITY STRESS

Since NaCl is the principal soil salinity stress, a research focus has been the transport systems that are involved in utilization of Na^+ as an osmotic solute (Blumwald et al., 2000; Hasegawa et al., 2000; Niu et al., 1995). Extensive research revealed that intracellular Na^+ homeostasis and salt tolerance are modulated by Ca^{2+}, and high Na^+ negatively affects K^+ acquisition (Rains & Epstein, 1967). Na^+ competes with K^+ for uptake through common transport systems, and does this effectively since the Na^+ in saline environments is usually considerably greater than K^+. Ca^{2+} enhances K^+/Na^+ selective intracellular accumulation (Maathuis et al., 1996; Rains & Epstein, 1967). Various workers have defined many of the molecular entities that mediate Na^+ and K^+ homeostasis and given insight into the function of Ca^{2+} in the regulation of these transport systems.

Recently, the SOS stress-signaling pathway was identified to be a pivotal regulator of plant ion homeostasis and salt tolerance (Hasegawa et al., 2000; Sanders, 2000). This signaling pathway functionally resembles the yeast calcineurin cascade that controls Na^+ influx and efflux across the plasma membrane (Bressan et al., 1998). Expression of an activated form of calcineurin in yeast or plants enhances salt tolerance further linking the functional similarity between the calcineurin and the SOS pathways (Mendoza et al., 1996; Pardo et al., 1998). Little is known about the mechanistic entities that are responsible for Cl^- transport or the regulation of Cl^- homeostasis (Hedrich, 1994).

1.5 NUTRIENT STRESS AND PLANT RESPONSE

Plants require at least 17 essential nutrients to complete their life cycle. Except for carbon, hydrogen, and oxygen, other essential nutrients are mineral nutrients, which are mainly acquired from soils by roots. Under normal conditions, the natural soil is low in nutrient content, and the availability of most essential mineral nutrients is very low and hard to meet the demand of plants (Liang et al., 2013). In most natural soils, low and fluctuating availability of most mineral nutrients, especially for macronutrients severely limits crop growth and yield (Gojon et al., 2009). Nutrient stress is a complex phenomenon and may result from either by low levels of availability of the element or by the presence of excess concentrations. In some cases, the presence of one element in excess concentrations may induce the deficiency of another element. In this context, attempts were made to show the availability, functional aspects, and deficiency and toxicity symptoms of 14 elements essential for the survival of plants and other elements, which lead to phytotoxicity. Visual deficiency symptoms grant an important basis for estimating the nutritional status of the plant. Deficiency symptoms are the consequence of metabolic disturbances at various stages of plant growth. Nutrient deficiency symptoms in plants vary from species to species and from element to element. The general plant symptoms are yellowing of leaves, darker than normal green color, interveinal chlorosis, necrosis, and twisting of leaves. Toxic levels of metals in soils are attained by the metal bearing soluble constituents in the natural soil or by waste disposal practices in mining, industrial manufacture, and urban sewage (Brown & Jones, 1975). Generally, each metal causing phytotoxicity exposes certain characteristic symptoms. The most general symptoms are stunting of growth and chlorosis of leaves. The toxicity symptoms observed may be due to a specific toxicity of the metal to the crop or due to an antagonism with other nutrients. A gist of nutrient stress on various plants has been given in Table 1.1. Some of the major influences of soil nutrient deficiency on plant metabolites are discussed in the following subsections.

TABLE 1.1   Effect of Nutrient Stress on Various Plant Metabolites.

Nutrient	Effect on plant	Plant used	Reference
Excess N**	Increased ethylene production	*Brassica juncea*	Iqbal et al. (2015)
Excess P**	Increased chamanzulene, β-farenzn, and essential oil content	*Matricaria recutita*	Jeshni et al. (2015)
Excess K**	Enhanced activity of enzymes like superoxide dismutase (SOD), peroxidase (POX), and ascorbate POX (APX)	*Solanum lycopersicum*	Hernandez et al. (2012)
Excess S	Increased glutathione production	*Brassica juncea*	Fatma et al. (2014)
Excess Ca	High antioxidant activity and photosynthesis	*Nicotiana tabacum*	Tan et al. (2011)
Mg deficiency*	Low photosynthesis	*Morus alba*	Tewari et al. (2006)

TABLE 1.1 *(Continued)*

Nutrient	Effect on plant	Plant used	Reference
Fe deficiency*	Enhanced production of phenols	*Parietaria judaica*	Tato et al. (2013)
Excess Zn	Increased malondialdehyde (MDA) level and low chlorophyll content	*Solanum lycopersicum*	Cherif et al. (2010)
Excess Ni	Enhanced synthesis of brassinosteroids (BRs)	*Brassica juncea*	Kanwar et al. (2012)

*Deficient means less than the optimum level (e.g., low levels for Mg and Fe are <50 mg kg^{-1} and <30 µg kg^{-1} soil, respectively).

**Excess means more than the optimum level (e.g., excess levels for N, P, K are >560, >28, and >280 kg/ha, respectively).

1.5.1 EFFECT OF N-DEFICIENCY ON PLANT PHENOLIC COMPOUNDS

Plant growth and development depends on an adequate supply of nitrogen (N) in order to synthesize the AAs, proteins, and nucleic acids (Sanchez et al., 2004). In addition, it is of great importance in the biochemistry of compounds such as enzymes, pigments, secondary metabolites, and polyamines (Maathuis, 2009). N deficiency leads to wide reshuffling of primary and secondary metabolism (Scheible et al., 2004), and low N has a drastic impact on the overall plant metabolism, inducing a shift from N-based to C-based compounds. Phenolic compounds are C-rich metabolites that represent the largest group of plant secondary metabolites which are generally synthesized through the shikimate pathway (Dixon & Paiva, 1995). The pathway continues, producing the AA aromatic phenylalanine, which is afterward deaminated by the enzyme phenylalanine ammonia lyase (PAL), the key enzyme in phenolic biosynthesis. PAL catalyses the non-oxidative deamination of L-phenylalanine to form cinnamic *trans*-acid. Phenolic compounds are further degraded oxidatively mainly by polyphenol oxidase and also by POXs. Rubio-Wilhelmi et al. (2012) reported a significant increase in tobacco plant phenolic compounds with increasing N deficiency. The increase in phenolic compound under low N is mainly attributed to the enhanced PAL activity (Kovácik et al., 2007). The first step of the shikimate pathway has been illustrated in Figure 1.3.

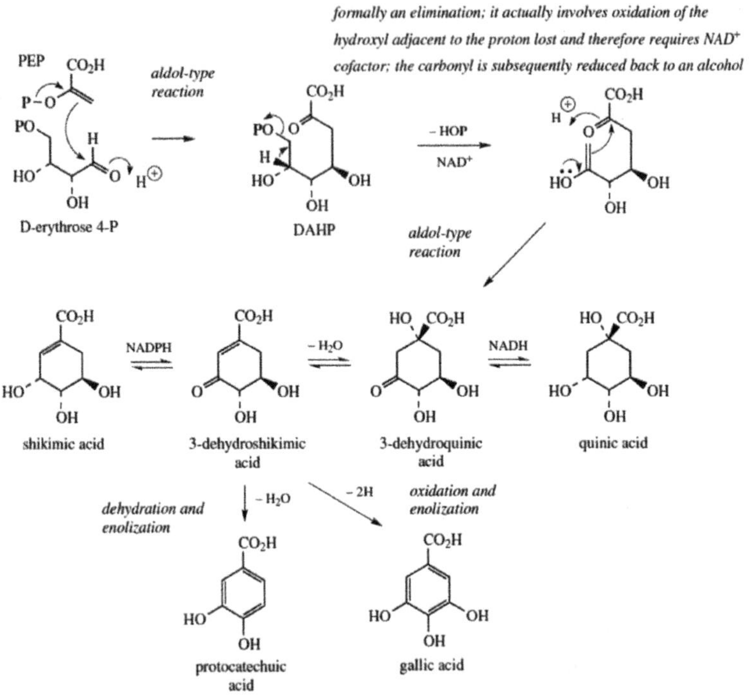
formally an elimination; it actually involves oxidation of the hydroxyl adjacent to the proton lost and therefore requires NAD+ cofactor; the carbonyl is subsequently reduced back to an alcohol
PEP
aldol-type reaction
D-erythrose 4-P
DAHP
- HOP
NAD+
aldol-type reaction
NADPH
- H2O
NADH
shikimic acid
3-dehydroshikimic acid
3-dehydroquinic acid
quinic acid
dehydration and enolization
- H2O
- 2H
oxidation and enolization
protocatechuic acid
gallic acid

FIGURE 1.3 Schematic flow chart of shikimate pathway in plants (from Dewick, P. M., Medicinal Natural Products, 3rd ed., John Wiley & Sons Ltd: Chichester, 2009. Reprinted with permission).

1.5.2 EFFECT OF N AND P DEFICIENCY ON PLANT ENZYMES

Two major mineral nutrients for plant growth and development are N and phosphorus (P) (Cai et al., 2013), which are known to be important components of key molecules in the cell (ATP, nucleic acids, chlorophyll, and phospholipids) and help modulate the expression of metabolic genes in rice (Sun et al., 2013). Furthermore, deficiencies in N and P result in an accumulation of carbohydrates in the leaves and roots and modify the shoot-to-root biomass ratio (Nielsen et al., 1998). Interestingly, genes related to photosynthesis and sucrose synthesis have been revealed to change their expression following P deficiency, indicating their involvement in the observed increase in growth (Cai et al., 2013). Moreover, it is well known that AGPase and soluble starch synthase play important roles in starch synthesis (Slattery et al., 2000). AGPase is a key regulatory enzyme that catalyzes the rate-limiting step of starch biosynthesis (Sakulsingharoj et al., 2004).

1.5.3 EFFECT OF K DEFICIENCY ON PLANT ANTIOXIDANTS

Potassium (K) is an essential macronutrient and plays an important role in metabolism as it functions as a cofactor of many enzymes and is required for charge balance and transport of metabolites (Marschner, 1995). Several lines of evidence show that K deficiency causes a decrease in sucrose export from source leaves (Mengel & Viro, 1974; Cakmak, 2005). Thus, the impairment in photosynthetic CO_2 fixation and decrease in sucrose export in K-deficient leaves could lead to enhanced oxygen photoreduction in the chloroplast via the Mehler reaction resulting in the production of ROS. In order to detoxify ROS, increases in the activities/contents of antioxidants are expected in leaves of K-deficient plants. Indeed, enhancement in the activities of antioxidant enzymes has been demonstrated in K-deficient bean leaves (Cakmak, 1994, 2005). Ding et al. (2008) also reported that the activities of SOD, CAT, and POX in the leaves of rice plants supplied with low K were higher than those supplied with high K.

1.5.4 EFFECT OF Fe DEFICIENCY ON PLANT PIGMENTS

Iron (Fe) deficiency is a major problem while growing high-value food crops. Fe deficient crops are characterized by the development of a pronounced interveinal chlorosis similar to that caused by Mg deficiency but occurring first on the youngest leaves. Interveinal chlorosis is sometimes followed by the chlorosis of the veins, so the whole leaf then becomes yellow or leading to necrotic leaves in severe cases. Fe deficiency lowers the formation of photosynthates (Bertamini et al., 2001). The inhibition of chlorophyll formation is partly due to the impaired protein synthesis. As the severity of Fe increases, the protein content per leaf area, the leaf cell volume, and the number of chloroplasts remain unaffected, whereas the chloroplast volume and the amount of protein per chloroplast decline (Terry, 1983). However, not all photosynthetic pigments are decreased to the same extent by Fe deficiency, xanthophylls being less affected than chlorophyll and β-carotene (Morales et al., 1990, 1994).

1.5.5   EFFECT OF N DEFICIENCY ON PLANT HORMONES

N is a macronutrient present in many key biological molecules and there-fore constitutes a limiting factor in agricultural systems. Avery et al. (1937) showed that extractable auxin was barely detectable in the tips of stems of N starved plants. Moreover, they determined that there is a significant correlation between NO_3^- and auxin content in leaves, for plants grown under different NO_3^- regimes. A 10-fold change in NO_3^- supply induces a 4-fold change in auxin concentration. For high NO_3^- concentrations, the auxin levels tend to plateau. Similarly, cytokinin content is also under the control of N supply (Sakakibara et al., 2006; Kiba et al., 2011). For instance, when tomato (*Solanum lycopersicum*) plants are grown with NH_4^+ as the sole N source, shoot growth is strongly reduced. At the molecular level, the *Arabidopsis ipt3* gene (an isopentenyl transferase responsible for the limiting step in the biosynthesis of cytokinins) is strongly induced by NO_3^- in both roots and shoots. Recent investigations have also revealed ethylene as a potential target of N-related signals. Indeed, ethylene production is enhanced in plants transferred from a low to a high NO_3^- containing media within 1 h of transfer (Tian et al., 2009).

1.6   PLANT SECONDARY METABOLITES AND STRESS

The secondary metabolites donot have a direct function in growth and maintenance of the cells, but have important ecological and protective functions against different forms of stress. The secondary compounds are major components of the metabolism of an organism which are synthe-sized in specialized cells or tissue. These metabolites are more complex in structure and functioning than primary compounds. Plants produce secondary metabolites as a response to adverse environmental conditions or in particular developmental stages. For example, exposure to UV radia-tion induces the biosynthesis of UV-absorbing compounds. Their biosyn-thesis starts from some primary metabolite or from intermediates of the primary metabolism and many of them accumulate in surprisingly high concentrations in some species. These chemicals are extremely diverse; many thousands have been identified in several major classes.

Stress is considered to be a significant deviation from the optimal conditions for living and functioning of plants which elicits changes and responses at all functional levels of the plant. Although the various impacts of plant stresses on secondary metabolism has been discussed in the previous sections of this chapter, this section will just give a concise account of the influence of environmental stress on certain specific secondary metabolites.

1.6.1 PHENOLS

Phenolic compounds are complex aromatic structural moieties which form the major component of plant secondary metabolite system. These substances are mostly produced and accumulated during various incidences of stress like water, nutrient, UV radiation, herbivory or pest infestation, and so on. Reports reveal that water stress induces the elevated synthesis and accumulation of *oleuropein* in *Olea europaea* (Petridis et al., 2012). High degree of plant wounding also influence the higher production of plant phenols. De Matos Nunes (2014) reported higher production of *uliginosin B* in *Hypericum polyanthemum* leaves. Higher salt concentration greatly affects the synthesis of *oleuropein* and *hydroxysterol* in *Olea europaea* (Petridis et al., 2012). A comprehensive list of important plant phenols synthesized under stress conditions has been provided in Table 1.2.

TABLE 1.2 Different Phenols Expressed during Plant Stress Conditions.

Environmental stress	Phenolic compounds	Plants	Reference
High water stress	Caffeoylquinic acid, chlorogenic acid, cryptochlorogenic acid, caffeoyl-hexoside, *p*-coumaroyl-hexoside, feruloyl-hexoside, caffeoyl-hexoside isomer, sinapoyl-hexoside	*Lycopersicon esculentum*	Sánchez-Rodríguez et al. (2012)
High heavy metal stress	Ellagic acid, vanillic acid, cinnamic acid derivate, *m*-coumaric acid, caffeic acid, caffeic acid derivate, *p*-coumaric acid derivate	*Erica andevalensis*	Márquez-García et al. (2009)
High UV-B	Chlorogenic acid, rutin	*Vaccinium corymbosum*	Inostroza-Blancheteau et al. (2014)

TABLE 1.2 *(Continued)*

Environmental stress	Phenolic compounds	Plants	Reference
Ca deficiency*	Chlorogenic acid, caffeic acid	*Solanum tuberosum*	Ngadze et al. (2014)
High salinity	Catechin, epicatechin, epicatechin benzyl thioether, hydroxycinnamic acid	*Mesembryanthemum edule*	Falleh et al. (2012)

*Deficient means less than the optimum level (for example: low level for Ca is <0.5 meq 100g⁻¹ soil).

*Deficient means less than the optimum level (for example: low level for Ca is <0.5 meq $100g^{-1}$ soil).

1.6.2 ALKALOIDS

Environmental stresses trigger the synthesis of various alkaloids in plants. Terpenoid indole alkaloids like *vinblastine* and *vincristine* are potent candidates in cancer therapy, produced during the events of high UV-B radiation (Binder et al., 2009). Do Nascimento et al. (2013) reported the synthesis of *brachycerine*, an antioxidant glucosidic indole alkaloid under osmotic and heavy metal stress in leaves *of Psychotria brachyceras*. Under the influence of heavy cold stress, *ergine* and *ergonovine* alkaloids are produced in endophyte-infected *Festuca sinensis* (Zhou et al., 2015).

1.6.3 TANNINS

Tannins are polyphenolic compounds with high molecular weights ranging from 500 to over 20,000. Environmental factors like nutrient availability, drought, salinity, pH, herbivory, ozone, and CO_2 concentration influence the concentration of tannin. Tannins are of two broad classes, namely condensed tannins and hydrolysable tannins. Tannins like *gallotannins* and *ellagitannins* are produced during high salinity (El-Lamey, 2012).

1.6.4 FLAVONOIDS

The concentration of flavonoids in plants varies with respect to soil nutrient status and climatic conditions. These complex aromatic compounds vary among the cultivars and plant phenological stage. Khan et al. (2011)

reported elevated concentration of *kaempferol* in *Brassica oleracea* under water stress conditions. Excessive light, water scarcity, and low soil K result in higher accumulation of *epigallocatechin-3-gallate, quercetin,* and *resveratrol* (Pirie et al., 2013). Higher synthesis of flavonoids: *2,3-dihydro-7-hydroxy-2-phenyl-4H-1-benzopyran-4-one, 2,3-dihydro-5,7-dihydroxy-2-phenyl-4H-1-benzopyran-4-one,* and *5,7-diacetoxy-2,3-dihydro-2-phenyl-4H-1-benzopyran-4-one* is reported under drought and low light stress (Daniels et al., 2015). Table 1.3 exhibits some of the major flavonoids synthesized under different stress events.

TABLE 1.3 Different Types of Flavonoids Expressed during Plant Stress Events.

Environmental stress	Flavonoid	Plant	Reference
High salinity	Naringenin, isoliquiritigenin, quercetin, umbelliferone, 7,4-dihydroxyflavone, hesperetin	*Phaseolus vulgaris*	Dardanelli et al. (2012)
Heavy irrigation	Rutin, quercetin, isoquercitin, isorhamnetin glucoside	*Vitis vinifera*	Zarrouk et al. (2012)
High UV-B	Kaempferol, quercetin, isorhamnetin glucoside, syringetin, myricetin, laricitrin	*Vitis vinifera*	Martínez-Lüscher et al. (2014)
High heat	Daidzin, daidzein, *m*-glycitin	*Capsicum annuum*	Khan et al. (2013)
Excess As*	Rutin, quercetin, hyperin, kaempferol	*Pteris* sp.	Wang et al. (2010)
High pathogenic infestation	Apigenin, chrysoeriol, tricin, luteolin, formononetin, medicarpin	*Medicago truncatula*	Jasinski et al. (2009)
High UV-C	Picied, quercetin	*Vitis vinifera*	Crupi et al. (2013)
Excess N & P*	Morin, rutin, quercetin, kaempferol, isorhamnetin	*Lycopersicon esculentum*	Stewart et al. (2001)

*Excess means more than the optimum level (e.g., excess levels for N and P are >560 and >28, respectively).

1.6.5 HORMONES

Plants control every kind of stress by specific hormones which allow defense responses against defined environmental conditions. Under heavy salt stress, plant hormones like *gibberellic acid* and *salicyclic acid* decreased while *abscisic acid* and *jasmonic acid* recorded a sharp increase (Hamayun et al., 2010). A similar experiment conducted by Babu et al. (2012) revealed a prominent increase in *indole acetic acid* and *abscisic acid* levels with elevating salt concentration. An interesting report by Koshita and Takahara (2004) showed that the concentration of gibberellic acid increases with salinity while *indole acetic acid* is mostly synthesized under moderate salinity conditions.

1.6.6   PIGMENTS

Plant pigments, except chlorophyll are mainly important due to their contribution in attracting pollinators. Various plant pigments like carotenoids, anthocyanins, xanthophylls, flavonoid pigments, and so on, perform different functions in a plant body. However, reports show that environmental stress poses prominent influence on the production of various plant pigments. Borghesi et al. (2011) showed that tomatoes grown in highly saline soils produced higher level of pigments like carotenoids and anthocyanins. Contrasting results were obtained by Doganlar et al. (2010) while growing tomato under salinity stress. Heavy metal stress (Cu^{2+}) influences the production of carotenoids in plants (Poonam et al., 2013).

1.6.7   ENZYMES

Various plant enzymes linked with oxidative stress management are up or down regulated under stress events. Weisany et al. (2012) conducted an experiment to ascertain the impact of salinity stress on enzyme production and activity in soybean. Results show that enzymes like *catalase*, *APX*, *polyphenoloxidase*, and *POX* increase with elevating salinity. Zhang et al. (2007) reported that higher metal concentration in soil influences the elevated production of *SOD*, *catalase*, and *POX*. Water stress depresses the production of *nitrate reductase* and *PAL* (Bardzik et al., 1970).

KEYWORDS

- terrestrial plants
- cellular organelles
- plant physiological system
- photosynthesis
- primary response

REFERENCES

Ahmad, P. Growth and Antioxidant Responses in Mustard (*Brassica juncea* L.) Plants Subjected to Combined Effect of Gibberellic Acid and Salinity. *Arch. Agron. Soil Sci.* **2010,** *56,* 575–588.

Ahmad, P.; Prasad, M. N. V. *Abiotic Stress Responses in Plants: Metabolism, Productivity and Sustainability,* Springer: New York, 2012.

Ahmad, P.; Umar, S. *Oxidative Stress: Role of Antioxidants in Plants,* Studium Press: New Delhi, 2011.

Al-Whaibi, M. H. Plant Heat-shock Proteins: A Mini Review. *J. King Saud Univ. Sci.* **2011,** *23,* 139–150.

Apel, K.; Hirt, H. Reactive Oxygen Species: Metabolism, Oxidative Stress, and Signal Transduction. *Ann. Rev. Plant Biol.* **2004,** *55,* 373–399.

Atkin, O. K.; Macherel, D. The Crucial Role of Plant Mitochondria in Orchestrating Drought Tolerance. *Ann. Bot.* **2009,** *103,* 581–597.

Avery, G. S.; Burkholder, P. R.; Creighton, H. B. Nutrient Deficiencies and Growth Hormone Concentration in *Helianthus* and *Nicotiana. Am. J. Bot.* **1937,** *24,* 553–557.

Babu, M. A.; Singh, D.; Gothandam, K. M. The Effect of Salinity on Growth, Hormones and Mineral Elements in Leaf and Fruit of Tomato Cultivar PKM1. *J. Animal Plant Sci.* **2012,** *22,* 159–164.

Bardzik, J. M.; Marsh, H. V.; Havis, J. R. Effects of Water Stress on the Activities of Three Enzymes in Maize Seedlings. *Plant Physiol.* **1970,** *47,* 828–831.

Berry, J. A.; Raison, J. K. *Responses of macrophytes to temperature.* In *Encyclopedia of Plant Physiology, New Series*; Lange, O. L., Nobel, P. S., Osmond, C. B., Ziegler, H., Eds.; Springer-Verlag: Berlin-Heidelberg, 1981; pp 277–338.

Berry, J. A.; Björkman, O. Photosynthetic Response and Adaptation to Temperature in Higher Plants. *Annu. Rev. Plant Physiol.* **1980,** *31,* 491–543.

Bertamini, M.; Nedunchezhian, N.; Borghi, B. Effect of Iron Deficiency Induced Changes on Photosynthetic Pigments, Ribulose-1,5-bisphosphate Carboxylase, and Photosystem Activities in Field Grown Grape Vine (*Vitis vinifera* L. cv. Pinot Noir) Leaves. *Photosynthesis* **2001,** *39,* 59–65.

Binder, B. Y. K.; Peebles, C. A. M.; Shanks, J. V.; San, K.-Y. The Effects of UV-B Stress on the Production of Terpenoid Indole Alkaloids in *Catharanthus roseus* Hairy Roots. *Biotechnol. Prog.* **2009**, *25*, 861–865.

Bita, C. E.; Gerats, T. Plant Tolerance to High Temperature in a Changing Environment: Scientific Fundamentals and Production of Heat Stress-tolerant Crops. *Front. Plant Sci.* **2013**. http://dx.doi.org/10.3389/fpls.2013.00273.

Blum, A. Drought Resistance, Water-Use Efficiency, and Yield Potential—Are They Compatible, Dissonant, or Mutually Exclusive? *Aust. J. Agric. Res.* **2005**, *56*, 1159–1168.

Blumwald, E.; Aharon, G. S.; Apse, M. P. Sodium Transport in Plant Cells. *Biochim. Biophys. Acta* **2000**, *1465*, 140–151.

Bohnert, H. J.; Jensen, R. G. Metabolic Engineering for Increased Salt Tolerance–the Next Step. *Aust. J. Plant Physiol.* **1996**, *23*, 661–667.

Bohnert, H. J.; Nelson, D. E.; Jensen, R. G. Adaptations to Environmental Stresses. *Plant Cell* **1995**, *7*, 1099–1111.

Borghesi, E.; Gonzalez-Miret, M. L.; Escudero-Gilete, M. L.; Malorgio, F.; Heredia, F. J.; Melendez-Martínez, A. J. Effects of Salinity Stress on Carotenoids, Anthocyanins, and Color of Diverse Tomato Genotypes. *J. Agric. Food Chem.* **2011**, *59*, 11676–11682.

Bressan, R. A.; Hasegawa, P. M.; Pardo, J. M. Plants Use Calcium to Resolve Salt Stress. *Trends Plant Sci.* **1998**, *3*, 411–412.

Brown, J. C; Jones, W. E. Heavy Metal Toxicity in Plants. I. A Crises in Embryo. *Commun. Soil Sci. Plant Anal.* **1975**, *6*, 421–438.

Cai, H. M.; Xie, W. B.; Lian, X. M. Comparative Analysis of Differentially Expressed Genes in Rice under Nitrogen and Phosphorus Starvation Stress Conditions. *Plant Mol. Biol. Report.* **2013**, *31*, 160–173.

Cajánek, M.; Štroch, M.; Lachetová, K; Kalina, J.; Špunda, V. Characterization of the Photosystem II Inactivation of Heat-stressed Barley Leaves as Monitored by the Various Parameters of Chlorophyll a Fluorescence and Delayed Fluorescence. *J. Photochem. Photobiol.* **1998**, *47*, 39–45.

Cakmak, I. Activity of Ascorbate-Dependent H_2O_2-Scavenging Leaves, but not in Phosphorus-Deficient Leaves. *J. Exp. Bot.* **1994**, *45*, 1259–1266.

Cakmak, I. The Role of Potassium in Alleviating Detrimental Effects of Abiotic Stress in Plants. *J. Plant Nutr. Soil Sci.* **2005**, *168*, 521–530.

Campbell, G. S.; Norman, J. M. *An Introduction to Environmental Biophysics.* 2nd ed.; Springer-Verlag: Berlin-Heidelberg.

Chaves, M. M.; Flexas, J.; Pinheiro, C. Photosynthesis under Drought and Salt Stress: Regulation Mechanisms from Whole Plant to Cell. *Ann. Bot.* **2009**, *103*, 551–560.

Cherif, J.; Derbel, N.; Nakkach, M.; von Bergmann, H.; Jemal, F.; Lakhdar, Z. B. Analysis of in vivo Chlorophyll Fluorescence Spectra to Monitor Physiological State of Tomato Plants Growing under Zinc Stress. *J. Phytochem. Photobiol. B: Biol.* **2010**, *101*, 332–339.

Cowling, S. A.; Sage, R. F. Interactive Effects of Low Atmospheric CO_2 and Elevated Temperature on Growth, Photosynthesis and Respiration in *Phaseolus vulgaris*. *Plant Cell Environ.* **1998**, *21*, 427–435.

Crupi, P.; Pichierri, A.; Basile, T.; Antonacci, D. Postharvest Stilbenes and Flavonoids Enrichment of Table Grape cv Redglobe (*Vitis vinifera* L.) as Affected by Interactive UV-C Exposure and Storage Conditions. *Food Chem.* **2013**, *141*, 802–808.

DaCosta, M.; Huang, B. Osmotic Adjustment Associated with Variation in Bent Grass Tolerance to Drought Stress. *J. Amer. Soc. Hort. Sci.* **2006**, *131*, 338–344.

Daniels, C. W.; Rautenbach, F.; Marnewick, J. L.; Valentine, A. J.; Babajide, O. J.; Mabusela, W. T. Environmental Stress Effect on the Phytochemistry and Antioxidant Activity of a South African Bulbous Geophyte, *Gethyllis multifolia* L. Bolus. *South Afr. J. Bot.* **2015**, *96*, 29–36.

Dardanelli, M. S.; De Cordoba, F. J. F.; Estevez, J.; Contreras, R.; Cubo, M. T.; Rodriguez-Carvajal, M. A.; Gil-Serrano, A. M.; Lopez-Baena, F. J.; Bellogin, R.; Manyani, H.; Ollero, F. J.; Megias, M. Changes in Flavonoids Secreted by *Phaseolus vulgaris* Roots in the Presence of Salt and the Plant Growth-promoting Rhizobacterium *Chryseobacterium balustinum*. *App. Soil Ecol.* **2012**, *57*, 31–38.

De Matos Nunes, J.; Bertodo, L. O. O.; Da Rosa, L. M. G.; Von Poser, G. L.; Rech, S. B. Stress Induction of Valuable Secondary Metabolites in *Hypericum polyanthemum* Acclimatized Plants. *South Afr. J. Bot.* **2014**, *94*, 182–189.

Demmig-Adams, B.; Adams, W. W. Photoprotection in an Ecological Context: The Remarkable Complexity of Thermal Energy Dissipation. *New Phytol.* **2006**, *171*, 11–21.

Desikan, R.; Hancock, J. T.; Neill, S. J. Regulation of the *Arabidopsis* Transcriptome by Oxidative Stress. *Plant Physiol.* **2001**, *127*, 159–172.

Dewick, P. M. The Shikimate Pathway: Aromatic Acids and Phenylpropanoids. In *Medicinal Natural Products: A Biosynthetic Approach*, 3rd ed., John Wiley & Sons: Chichester, UK, 2009. http://dx.doi.org/10.1002/9780470742761.ch4.

Ding, Y.-C.; Chang, C.-R.; Luo, W.; Wu, Y.-S.; Ren, X.-L.; Wang, P.; Xu, G.-H. High Potassium Aggravates the Oxidative Stress Induced by Magnesium Deficiency in Rice Leaves. *Pedosphere* **2008**, *18*, 316–327.

Dixon, R. A.; Paiva, N. L. Stress-induced Phenylpropanoid Metabolism. *Plant Cell* **1995**, *7*, 1085–1097.

Do Nascimento, N. C.; Menguer, P. K.; Henriques, A. T.; Fett-Neto, A. G. Accumulation of Brachycerine, an Antioxidant Glucosidic Indole Alkaloid, is Induced by Abscisic Acid, Heavy Metal, and Osmotic Stress in Leaves of *Psychotria brachyceras*. *Plant Physiol. Biochem.* **2013**, *73*, 33–40.

Dobson, C. M.; Sali, A.; Karplus, M. Protein Folding: A Perspective from Theory and Experiment. *Angew. Chem. Int. Ed.* **1998**, *37*, 868–893.

Doganlar, Z. B.; Demir, K.; Basak, H.; Gul, I. Effects of Salt Stress on Pigment and Total Soluble Protein Contents of Three Different Tomato Cultivars. *Afr. J. Agric. Res.* **2010**, *5*, 2056–2065.

El-Lamey, T. M. Effect of Salinity on Tannins Content of *Leucaena leucocephala* (Lam.) de Wit. and *Prosopis chilensis* (Molina) Stuntz and Techniques for their Reduction. *Egypt. J. Bot.* **2012**, 51–63.

Enami, I.; Tomo, T.; Kitamura, M.; Katoh, S. Immobilization of the 3 Extrinsic Proteins in Spinach Oxygen-evolving Photosystem-II Membranes—Roles of the Proteins in Stabilization of Binding of Mn and Ca^{2+}. *Biochim. Biophys. Acta* **1994**, *1185*, 75–80.

Fait, A. ; Fromm, H.; Walter, D.; Galili, G.; Fernie, A. Highway or Byway: The Metabolic Role of the GABA Shunt in Plants. *Trends Plant Sci.* **2007**, *13*, 14–19.

Falleh, H.; Jalleli, I.; Ksouri, R.; Boulaaba, M.; Guyot, S.; Magne, C.; Abdelly, C. Effect of Salt Treatment on Phenolic Compounds and Antioxidant Activity of Two *Mesembryanthemum edule* Provenances. *Plant Physiol. Biochem.* **2012**, *52*, 1–8.

Fatma, M.; Asgher, M.; Masood, A.; Khan, N. A. Excess Sulfur Supplementation Improves Photosynthesis and Growth in Mustard under Salt Stress through Increased Production of Glutathione. *Environ. Exp. Bot.* **2014,** *107,* 55–63.

Feder, M. E. Integrative Biology of Stress: Molecular Actors, the Ecological Theater, and the Evolutionary Play, Proceedings of the International Symposium on Environmental Factors, Cellular Stress and Evolution, Varanasi, India, October 13–15, 2006.

Florez-Sarasa, I. D.; Bouma, T. J.; Medrano, H.; Azcon-Bieto, J.; Ribas-Carbo, M. Contribution of the Cytochrome and Alternative Pathways to Growth Respiration and Maintenance Respiration in *Arabidopsis thaliana. Physiol. Plant.* **2007,** *129,* 143–151.

Goicoechea, N.; Dolezal, K.; Antolin, M. C.; Stmad, M.; Sanchez- Diaz, M. Influence of Mycorrhizae and Rhizobium on Cytokinin Content in Drought-stressed Alfalfa. *J. Exp. Bot.* **1995,** *46,* 1543–1549.

Gojon, A.; Nacry, P.; Davidian, J. C. Root Uptake Regulation: A Central Process for NPS Homeostasis in Plants. *Curr. Opin. Plant Biol.* **2009,** *12,* 328–338.

Gombos, Z.; Wada, H.; Hideg, E.; Murata, N. The Unsaturation of Membrane Lipids Stabilizes Photosynthesis Against Heat Stress. *Plant Physiol.* **1994,** *104,* 563–567.

Gonzalez-Meler, M. A.; Giles, L.; Thomas, R. B.; Siedow, J. N. Metabolic Regulation of Leaf Respiration and Alternative Pathway Activity in Response to Phosphate Supply. *Plant Cell Environ.* **2001,** *24,* 205–215.

Guilioni, L.; Wéry, J.; Lecoeur, J. High Temperature and Water Deficit may Reduce Seed Number in Field Pea Purely by Decreasing Plant Growth Rate. *Funct. Plant Biol.* **2003,** *30,* 1151–1164.

Gupta, S. C.; Sharma, A.; Mishra, M.; Mishra, R.; Chowdhuri, D. K. Heat Shock Proteins in Toxicology: How Close and How Far? *Life Sci.* **2010,** *86,* 377–384.

Hamayun, M.; Khan, S. A.; Khan, A. L.; Shinwari, Z. K.; Hussain, J.; Sohn, E.-Y.; Kang, S.-M.; Kim, Y.-H.; Khan, M. A.; Lee, I.-J. Effect of Salt Stress on Growth Attributes and Endogenous Growth Hormones of Soybean Cultivar Hwangkeumkong. *Pak. J. Bot.* **2010,** *42,* 3103–3112.

Hasegawa, P. M.; Bressan, R. A.; Zhu, J.-K.; Bohnert, H. J. Plant Cellular and Molecular Responses to High Salinity. *Annu. Rev. Plant Physiol. Plant Mol. Biol.* **2000,** *51,* 463–499.

Hedrich, R. Voltage-Dependent Chloride Channels in Plant Cells: Identification, Characterization, and Regulation of Guard Cell Anion Channel. *Curr. Topics Membr.* **1994,** *42,* 1–33.

Hernandez, M.; Fernandez-Gracia, N.; Garcia-Garma, J.; Rubio-Asensio, J. S.; Rubio, F.; Olmos, E. Potassium Starvation Induces Oxidative Stress in *Solanum lycopersicum* L. Roots. *J. Plant Physiol.* **2012,** *169,* 1366–1374.

Hong, S. W.; Lee, U.; Vierling, E. Arabidopsis Hot Mutants Define Multiple Functions Required for Acclimation to High Temperatures. *Plant Physiol.* **2003,** *132,* 757–767.

Hong, S. W.; Vierling, E. Mutants of *Arabidopsis thaliana* Defective in the Acquisition of Tolerance to High Temperature Stress. *Proc. Nat. Acad. Sci.* **2000,** *97,* 4392–4400.

Hsiao, T. C.; Xu , L. K. Sensitivity of Growth of Roots Versus Leaves to Water Stress: Biophysical Analysis and Relation to Water Transport. *J. Exp. Bot.* **2000,** *51,* 1595–1616.

Inostroza-Blancheteau, C.; Reyes-Díaz, M.; Arellano, A.; Latsague, M.; Acevedo, P.; Loyola, R.; Arce-Johnson, P.; Alberdi, M. Effects of UV-B Radiation on Anatomical Characteristics, Phenolic Compounds and Gene Expression of the Phenylpropanoid Pathway in Highbush Blueberry Leaves. *Plant Physiol. Biochem.* **2014,** *85,* 85–95.

Iqbal, N.; Umar, S.; Khan, N. A. Nitrogen Availability Regulates Proline and Ethylene Production and Alleviates Salinity Stress in Mustard (*Brassica juncea*). *J. Plant Physiol.* **2015**. http://dx.doi.org/10.1016/j.jplph.2015.02.006.

Iuchi, S.; Kobayashi, M.; Taji, T.; Naramoto, M.; Seki, M.; Kato, T.; Tabata, S.; Kakubari, Y.; Yamaguchi-Shinozaki, K.; Shinozaki, K. Regulation of Drought Tolerance by Gene Manipulation of 9-cisepoxycarotenoid Dioxygenase, a Key Enzyme in Abscisic Acid Biosynthesis in Arabidopsis. *Plant J.* **2001**, *27*, 325–333.

Jager, E.; Symons, G. M.; Ross, J. J.; Reid, J. B. Do Brassinosteroids Mediate the Water Stress Response? *Physiol. Plant.* **2008**, *133*, 417–425.

James, R. A.; Blake, C.; Byrt, C. S.; Munns, R. Major Genes for Na+ Exclusion, Nax1 and Nax2 (wheatHKT1;4 and HKT1;5), Decrease Na+ Accumulation in Bread Wheat Leaves under Saline and Waterlogged Conditions. *J. Exp. Bot.* **2011**, *62*, 2939–2947.

Jasinski, M.; Kachlicki, P.; Rodziewicz, P.; Figlerowicz, M.; Stobiecki, M. Changes in the Profile of Flavonoid Accumulation in *Medicago truncatula* Leaves during Infection with Fungal Pathogen *Phoma medicaginis*. *Plant Physiol. Biochem.* **2009**, *47*, 847–853.

Jeshni, M. G.; Mousavinik, M.; Khammari, I.; Rahimi, M. The Changes of Yield and Essential Oil Components of German Chamomile (*Matricaria recutita* L.) under Application of Phosphorus and Zinc Fertilizers and Drought Stress Conditions. *J. Saudi Soc. Agric. Sci.* **2015**. http://dx.doi.org/10.1016/j.jssas.2015.02.003.

Kagale, S.; Divi, U. K.; Krochko, J. E.; Keller, W. A.; Krishna, P. Brassinosteroid Confers Tolerance in *Arabidopsis thaliana* and *Brassica napus* to a Range of Abiotic Stresses. *Planta* **2007**, *225*, 353–364.

Kanwar, M. K.; Bhardwaj, R.; Arora, P.; Chowdhary, S. K.; Sharma, P.; Kumar, S. Plant Steroid Hormones Produced under Ni Stress are Involved in the Regulation of Metal Uptake and Oxidative Stress in *Brassica juncea* L. *Chemosphere* **2012**, *86*, 41–49.

Khan, A. L.; Kang, S.-M.; Dhakal, K. H.; Hussain, J.; Adnan, M.; Kim, J.-G.; Lee, I.-J. Flavonoids and Amino Acid Regulation in *Capsicum annuum* L. by Endophytic Fungi under Different Heat Stress Regimes. *Sci. Hortic.* **2013**, *155*, 1–7.

Khan, M. A. M.; Ulrichs, C.; Mewis, I. Effect of Water Stress and Aphid Herbivory on Flavonoids in Brocolli (*Brassica oleracea* var. *italica* Plenk). *J. App. Bot. Food Qual.* **2011**, *84*, 178–182.

Kiba, T.; Kudo, T.; Kojima, M.; Sakakibara, H. Hormonal Control of Nitrogen Acquisition: Roles of Auxin, Abscisic Acid, and Cytokinin. *J. Exp. Bot.* **2011**, *62*, 1399–1409.

Koshita, Y.; Takahara, T. Effect of Water Stress on Flower-bud Formation and Plant Hormone Content of Satsuma Mandarin (*Citrus unshiu* Marc.). *Sci. Hortic.* **2004**, *99*, 301–307.

Kosova, K.; Vitamvas, P.; Prasil, I. T.; Renaut, J. Plant Proteome Changes under Abiotic Stress Contribution of Proteomics Studies to Understanding Plant Stress Response. *J. Proteomics* **2011**, *74*, 1301–1322.

Kovácik, J.; Klejdus, B.; Backor, M.; Repcák, M. Phenylalanine Ammonia-lyase Activity and Phenolic Compounds Accumulation in Nitrogen-Deficient *Matricaria chamomilla* leaf Rosettes. *Plant Sci.* **2007**, *172*, 393–399.

Krasensky, J.; Jonak, C. Drought, Salt, and Temperature Stress-induced Metabolic Rearrangements and Regulatory Networks. *J. Exp. Bot.* **2012**. http://dx.doi.org/10.1093/jxb/err460.

Lambers, H.; Robinson, S. A.; Ribas-Carbo, M. Regulation of Respiration in vivo. In *Plant Respiration: From Cell to Ecosystem. Advances in Photosynthesis and Respiration Series*; Lambers, H., Ribas-Carbo, M., Eds.; Springer: The Netherlands, 2005; Vol. 18; pp 1–15.

Lawlor, D. W.; Cornic, G. Photosynthetic Carbon Assimilation and Associated Metabolism in Relation to Water Deficits in Higher Plants. *Plant Cell Environ.* **2002**, *25*, 275–294.

Lawlor, D. W.; Tezara, W. Causes of Decreased Photosynthetic Rate and Metabolic Capacity in Water-deficient Leaf Cells: A Critical Evaluation of Mechanisms and Integration of Processes. *Ann. Bot.* **2009**, *103*, 561–579.

Lei Y.; Yin C.; Li C. Differences in Some Morphological, Physiological, and Biochemical Responses to Drought Stress in Two Contrasting Populations of *Populus przewalskii*. *Physiologia Plantarum.* **2006**, *127*, 182–191.

Levitt, J. *Responses of Plants to Environmental Stresses*, 2nd ed.; Academic Press: New York, 1980, p 497.

Levitt, M.; Gerstein, M.; Huang, E.; Subbiah, S.; Tsai, J. Protein Folding: The Endgame. *Annu. Rev. Biochem.* **1997**, *66*, 549–579.

Liang, C.; Tian, J.; Liao, H. Proteomics Dissection of Plant Responses to Mineral Nutrient Deficiency. *Proteomics* **2013**, *13*, 624–636.

Lindquist, S.; Crig, E. A. The Heat-Shock Proteins. *Annu. Rev. Genet.* **1988**, *22*, 631–677.

Maathuis, F. J. M. Physiological Functions of Mineral Macronutrients. *Curr. Opin. Plant Biol.* **2009**, *12*, 250–325.

Maathuis, F. J. M.; Verlin, D.; Smith, F. A.; Sanders, D.; Ferneáßndez, J. A.; Walker, N. A. The Physiological Relevance of Na+-coupled K+-transport. *Plant Physiol.* **1996**, *112*, 1609–1616.

Mackerness, S. A. H.; John, C. F.; Jordan, B.; Thomas, B. Early Signalling Components in Ultraviolet-B Responses: Distinct Roles for Different Reactive Oxygen Species and Nitric Oxide. *FEBS Lett.* **2001**, *489*, 237–242.

Mahajan, S.; Tuteja, N. Cold, Salinity and Drought Stresses: An Overview. *Arch. Biochem. Biophys.* **2005**, *444*, 139–158.

Márquez-García, B.; Fernández, M. A.; Córdoba, F. Phenolics Composition in *Erica* sp. Differentially Exposed to Metal Pollution in the Iberian Southwestern Pyritic Belt. *Bioresour. Tech.* **2009**, *100*, 446–451.

Marschner, H. *Mineral Nutrition of Higher Plants*. Academic Press: San Diego, 1995.

Martínez-Lüscher, J.; Torres, N.; Hilbert, G.; Richard, T.; Sánchez-Díaz, M.; Delrot, S.; Aguirreolea, J.; Pascual, I.; Gomès, E. Ultraviolet-B Radiation Modifies the Quantitative and Qualitative Profile of Flavonoids And Amino Acids in Grape Berries. *Phytochemistry* **2014**, *102*, 106–114.

Mendoza, I.; Quintero, F. J.; Bressan, R. A.; Hasegawa, P. M.; Pardo, J. M. Activated Calcineurin Confers High Tolerance to Ion Stress and Alters the Budding Pattern and Cell Morphology of Yeast Cells. *J. Biol. Chem.* **1996**, *271*, 23061–23067.

Mengel, H.; Viro, M. Effect of Potassium Supply on the Transport of Photosynthates to the Fruits of Tomatoes (*Lycopersicon esculentum*). *Physiol. Plant* **1974**, *30*, 295–300.

Morales, F.; Abadia, A.; Abadia, J. Characterization of the Xanthophyll Cycle and Other Photosynthetic Pigments Changes Induced by Iron Deficiency in Sugar Beet (*Beta vulgaris* L.). *Plant Physiol.* **1990**, *94*, 607–613.

Morales, F.; Abadia, A.; Belkhodjia, R.; Abadia, J. Iron Deficiency-induced Changes in the Photosynthetic Pigment Composition of Field Grown Pear (*Pyrus communis* L.) Leaves. *Plant Cell Environ.* **1994**, *17*, 1153–1160.

Morimoto, R. I. Cells in Stress: The Transcriptional Activation of Heat Shock Genes. *Science* **1993**, *259*, 1409–1410.

Morimoto, R. I.; Santoro, M. G. Stress-Inducible Responses and Heat Shock Proteins: New Pharmacologic Targets for Cytoprotection. *Nat. Biotechnol.* **1998**, *16*, 833–838.

Morimoto, R. I.; Tissieres, A.; Georgopoulos, C. *Heat Shock Proteins: Structure, Function and Regulation.* Cold Spring Harbor Lab. Press: New York, 1994.

Munns, R. Genes and Salt Tolerance: Bringing them Together. *New Phytol.* **2005**, *167*, 645–663.

Munns, R.; Tester, M. Mechanisms of Salinity Tolerance. *Ann. Rev. Plant Biol.* **2008**, *59*, 651–681.

Nahar, K.; Hasanuzzaman, M.; Majumder, R. R. Effect of Low Temperature Stress in Transplanted Aman Rice Varieties Mediated by Different Transplanting Dates. *Acad. J. Plant Sci.* **2009**, *2*, 132–138.

Ngadze, E.; Coutinho, T. A.; Icishahayo, D.; Van der Waals, J. Effect of Calcium Soil Amendments on Phenolic Compounds and Soft Rot Resistance in Potato Tubers. *Crop Prot.* **2014**, *62*, 40–45.

Nielsen, T. H.; Krapp, A.; Roper, S. U.; Stitt, M. The Sugar-mediated Regulation of Genes Encoding the Small Subunit of Rubisco and the Regulatory Subunit of ADP Glucose Pyrophosphorylase is Modified by Phosphate and Nitrogen. *Plant Cell Environ.* **1998**, *21*, 443–454.

Niu, X.; Bressan, R. A.; Hasegawa, P. M.; Pardo J. M. Ion Homeostasis in NaCl Stress Environments. *Plant Physiol.* **1995**, *109*, 735–742.

Niyogi, K. K. Photoprotection Revisited: Genetic and Molecular Approaches. *Annu. Rev. Plant Physiol. Plant Mol. Biol.* **1999**, *50*, 333–359.

Ntatsi, G.; Savvas, D.; Ntatsi, G.; Klaring, H. –P.; Schwarz, D. Growth, Yield, and Metabolic Responses of Temperature-Stressed Tomato to Grafting onto Rootstocks Differing in Cold Tolerance. *J. Am. Soc. Hort. Sci.* **2014**, *139*, 230–243.

Nuccio, M. L.; Rhodes, D.; McNeil, S. D.; Hanson, A. D. Metabolic Engineering of Plants for Osmotic Stress Resistance. *Curr. Opin. Plant Biol.* **1999**, *2*, 128–134.

Pardo, J. M.; Reddy, M. P.; Yang, S.; Maggio, A.; Huh, G.-H.; Matsumoto, T.; Coca, M. A.; Paino-D'Urazo, M.; Koiwa, H.; Yun, D.-J.; Watad, A. A.; Bressan, R. A.; Hasegawa, P. M. Stress Signaling through Ca2+/Calmodulin-Dependent Protein Phosphatase Calcineurin Mediates Salt Adaptation in Plants. *Proc. Natl. Acad. Sci.* **1998**, *95*, 9681–9686.

Parent, B.; Hachez,C.; Redondo, E.; Simonneau, T.; Chaumont, F.; Tardieu, F. Drought and Abscisic Acid Effects on Aquaporin Content Translate into Changes in Hydraulic Conductivity and Leaf Growth Rate: A Trans-scale Approach. *Plant Physiol.* **2009**, *149*, 2000–2012.

Peleg, Z.; Blumwald, E. Hormone Balance and Abiotic Stress Tolerance in Crop Plants. *Curr. Opin. Plant Biol.* **2011**, *14*, 290–295.

Petridis, A.; Therios, I.; Samouris, G.; Koundouras, S.; Giannakoula, A. Effect of Water Deficit on Leaf Phenolic Composition, Gas Exchange, Oxidative Damage and Antioxidant Activity of Four Greek Olive (*Olea europaea* L.) Cultivars. *Plant Physiol. Biochem.* **2012**, *60*, 1–11.

Petridis, A.; Therios, I.; Samouris, G.; Tananaki, C. Salinity-induced Changes in Phenolic Compounds in Leaves and Roots of Four Olive Cultivars (*Olea europaea* L.) and Their Relationship to Antioxidant Activity. *Environ. Exp. Bot.* **2012**, *79*, 37–43.

Pirie, A.; Parsons, D.; Renggli, J.; Narkowicz, C.; Jacobson, G. A.; Shabala, S. Modulation of Flavonoid and Tannin Production of *Carpobrotus rossii* by Environmental Conditions. *Environ. Exp. Bot.* **2013**, *87*, 19–31.

Plaxton, W.; Podesta, F. The Functional Organization and Control of Plant Respiration. *Crit. Rev. Plant Sci.* **2006**, *25*, 159–198.

Poonam, S.; Kaur, H.; Geetika,S. Effect of Jasmonic Acid on Photosynthetic Pigments and Stress Markers in *Cajanus cajan* (L.) Millsp. Seedlings under Copper Stress. *Am. J. Plant Sci.* **2013**, *4*, 817–823.

Pospisilova, J.; Synkova, H.; Rulcova, J. Cytokinins and Water Stress. *Biol. Plant.* **2000**, *43*, 321–328.

Prasad, P. V. V.; Boote, K. J.; Allen, L. H.; Thomas, J. M. G. Effects of Elevated Temperature and Carbon Dioxide on Seed-set and Yield of Kidney Bean (*Phaseolus vulgaris* L.). *Global Change Biol.* **2002**, *8*, 710–721.

Queitsch, C.; Hong, S. W.; Vierling, E.; Lindquist, S. Heat Shock Protein 101 Plays a Crucial Role in Thermotolerance in Arabidopsis. *Plant Cell.* **2000**, *12*, 479–492.

Raghavendra, A. S.; Padmasree, K. Beneficial Interactions of Mitochondrial Metabolism with Photosynthetic Carbon Assimilation. *Trends Plant Sci.* **2003**, *8*, 546–553.

Rahnama, A.; James, R. A.; Poustini, K.; Munns, R. Stomatal Conductance as a Screen for Osmotic Stress Tolerance in Durum Wheat Growing in Saline Soil. *Func. Plant Biol.* **2010**, *37*, 255–263.

Rains, D.; Epstein, E. Sodium Absorption by Barley Roots. Its Mediation by Mechanisms 2 of Alkali Cation Transport. *Plant Physiol.* **1967**, *42*, 319–323.

Rhodes, D.; Hanson, A. D. Quaternary Ammonium and Tertiary Sulfonium Compounds in Higher Plants. *Ann. Rev. Plant Physiol. Plant Mol. Biol.* **1993**, *44*, 357–384.

Roitsch, T. Source-Sink Regulation by Sugar and Stress. *Curr. Opin. Plant Biol.* **1999**, *2*, 198–206.

Rouhier, N.; Vieira Dos Santos, C.; Tarrago, L.; Rey, P. Plant Methionine Sulfoxide Reductase A and B Multigenic Families. *Photosynth. Res.* **2006**, *89*, 247–262.

Rozema, J.; Flowers, T. Ecology: Crops for a Salinized World. *Science*, **2008**, *322*, 1478–1480.

Rubio-Wilhelmi, M. D. M.; Sanchez-Rodriguez, E.; Leyva, R.; Blasco, B.; Romero, L.; Blumwald, E.; Ruiz, J. M. Response of Carbon and Nitrogen-rich Metabolites to Nitrogen Deficiency in P_{SARK} ::IPT Tobacco Plants. *Plant Physiol. Biochem.* **2012**, *57*, 231–237.

Rurek, M. Plant Mitochondria under a Variety of Temperature Stress Conditions. *Mitochondrion.* **2014**, *19*, 289–294.

Sade, N.; Alem, G.; Moshelion, M. Risk-taking Plants: Anisohydric Behavior as a Stress-resistance Trait. *Plant Signal. Behav.* **2012**, *7*, 1–4.

Sairam, S. K.. Effects of Homobrassinolide Application on Plant Metabolism and Grain Yield Under Irrigated and Moisture-stress Conditions of Two Wheat Varieties. *Plant Growth Regul.* **1994**, *14*, 173–181.

Sakakibara, H.; Takei, K.; Hirose, N. Interactions between Nitrogen and Cytokinin in the Regulation of Metabolism and Development. *Trends Plant Sci.* **2006**, *11*, 440–448.

Sakulsingharoj, C.; Choi, S. B.; Hwang, S. K.; Edwards, G. E.; Bork, J.; Meyer, C. R.; Preiss, J. T.; Okita, W. Engineering Starch Biosynthesis for Increasing Rice Seed Weight: The Role of the Cytoplasmic ADP-Glucose Pyrophosphorylase. *Plant Sci.* **2004**, *167*, 1323–1333.

Sanchez, E.; Rivero, R. M.; Ruiz, J. M.; Romero, L. Changes in Biomass, Enzymatic Activity and Protein Concentration in Roots and Leaves of Green Bean Plants (*Phaseolus vulgaris* L. cv. Strike) under High NH_4NO_3 Application Rates. *Sci. Hortic.* **2004**, *99*, 237–248.

Sánchez-Rodríguez, E.; Ruiz, J. M.; Ferreres, F.; Moreno, D. A. Phenolic Profiles of Cherry Tomatoes as Influenced by Hydric Stress and Rootstock Technique. *Food Chem.* **2012**, *134*, 775–782.

Sanders, D. Plant Biology: The Salty Tale of *Arabidopsis*. *Curr. Biol.* **2000**, *10*, 486–488.

Santarius, K. A. Sites of Heat Sensitivity in Chloroplasts and Differential Inactivation of Cyclic and Noncyclic Photophosphorylation by Heating. *J. Therm. Biol.* **1975**, *1*, 101–107.

Santarius, K. A.; Müller, M. Investigations on Heat Resistance of Spinach Leaves. *Planta* **1979**, *146*, 529–538.

Scheible, W. R.; Morcuende, R.; Czechowski, T.; Frizt, C.; Osuna, D.; Palacios-Rojas, N.; Schindelash, D.; Thimm, O.; Udvardi, M. K.; Stitt, M. Genome-wide Reprogramming of Primary and Secondary Metabolism, Protein Synthesis, Cellular Growth Processes, and the Regulatory Infrastructure of Arabidopsis in Response to Nitrogen. *Plant Physiol.* **2004**, *136*, 2483–2499.

Schmitt, E.; Gehrmann, M.; Brunet, M.; Multhoff, G.; Garrido, C. Intracellular and Extracellular Functions of Heat Shock Proteins: Repercussions in Cancer Therapy. *J. Leukocyte Biol.* **2007**, *81*, 15–27.

Schuster, W. S.; Monson, R. K. An Examination of the Advantages of C_3–C_4 Intermediate Photosynthesis in Warm Environments. *Plant Cell Environ.* **1990**, *13*, 903–912.

Shao, H. B.; Chu, L. Y.; Jaleel, C. A.; Manivannan, P.; Panneerselvam, R.; Shao, M. A. Understanding Water Deficit Stress-induced Changes in the Basic Metabolism of Higher Plants-Biotechnologically and Sustainably Improving Agriculture and the Ecoenvironment in Arid Regions of the Globe. *Crit. Rev. Biotechnol.* **2009**, *29*, 131–151.

Shao, H. B.; Chu, L. Y.; Lu, Z. H.; Kang, C. M. Primary Antioxidant Free Radical Scavenging and Redox Signaling Pathways in Higher Plant Cells. *Int. J. Biol. Sci.* **2008**, *4*, 8–14.

Sharp, R. E. Interaction with Ethylene: Changing Views on the Role of Abscisic Acid in Root and Shoot Growth Responses to Water Stress. *Plant, Cell Environ.* **2002**, *25*, 211–222.

Siedow, J. N.; Umbach, A. L. The Mitochondrial Cyanideresistant Oxidase: Structural Conservation Amid Regulatory Diversity. *Biochim. Biophys. Acta* **2000**, *1459*, 432–439.

Slattery, C. J.; Kavakli, I. H.; Okita, T. W. Engineering Starch for Increased Quantity and Quality. *Trends Plant Sci.* **2000**, *5*, 291–298.

Stewart, A. J.; Chapman, W.; Jenkins, G. I.; Graham, I.; Martin, T.; Crozier, A. The Effect of Nitrogen and Phosphorus Deficiency on Flavonol Accumulation in Plant Tissues. *Plant Cell Environ.* **2001**, *24*, 1189–1197.

Sun, W.; Huang, A.; Sang, Y.; Fu, Y.; Yang, Z. Carbon–Nitrogen Interaction Modulates Plant Growth and Expression of Metabolic Genes in Rice. *J. Plant Growth Regul.* **2013**, *32*, 575–584.

Taira, M.; Valtersson, U.; Burkhardt, B.; Ludwig, R. A. *Arabidopsis thaliana* GLN2-Encoded Glutamine Synthetase is Dual Targeted to Leaf Mitochondria and Chloroplasts. *Plant Cell* **2004**, *16*, 2048–2058.

Taiz, L.; Zeiger, E. *Plant Physiology*. Sinauer Associates, Inc.: Sunderland, MA, 1998.

Tan, W.; Meng, Q. W.; Brestic, M.; Olsovska, K.; Yang, X. Photosynthesis is Improved by Exogenous Calcium in Heat-stressed Tobacco Plants. *J. Plant Physiol.* **2011**, *168*, 2063–2071.

Tato, L.; De Nisi, P.; Donnini, S.; Zocchi, G. Low Iron Availability and Phenolic Metabolism in a Wild Plant Species (*Parietaria judaica* L.). *Plant Physiol. Biochem.* **2013**, *72*, 145–153.

Terry, N. Limiting Factors in Photosynthesis: IV. Iron Stress Mediated Changes in Light-harvesting and Electron Transport Capacity and Its Effects on Photosynthesis *in vivo*. *Plant Physiol.* **1983**, *71*, 855–860.

Terzaghi, W. B.; Fork, D. C.; Berry, J. A.; Field, C. B. Low and High Temperature Limits to PSII. A Survey using *trans*-Parinaric Acid, Delayed Light Emission, and Fo Chlorophyll Fluorescence. *Plant Physiol.* **1989**, *91*, 1494–1500.

Tewari, R. K.; Kumar, P.; Sharma, P. N. Magnesium Deficiency Induced Oxidative Stress and Antioxidant Responses in Mulberry Plants. *Sci. Hortic.* **2006**, *108*, 7–14.

Thameur, A.; Ferchichi, A.; Lopez-Carbonell, M. Quantification of Free and Conjugated Abscisic Acid in Five Genotypes of Barley (*Hordeum vulgare* L.) under Water Stress Conditions. *S. Afr. J. Bot.* **2011**, *77*, 222–228.

Thompson, L. K.; Blaylock, R.; Sturtevant, J. M.; Brudvig, G. W. Molecular Basis of the Heat Denaturation of Photosystem II. *Biochemistry* **1989**, *28*, 6686–6695.

Tian, Q. Y.; Sun, P.; Zhang, W. H. Ethylene is Involved in Nitrate-dependent Root Growth and Branching in *Arabidopsis thaliana*. *New Phytol.* **2009**, *184*, 918–931.

Timperio, A. M.; Egidi, M. G.; Zolla, L. Proteomics Applied on Plant Abiotic Stresses: Role of Heat Shock Proteins (HSP). *J. Proteomics* **2008**, *71*, 391–411.

Van Aken, O.; Whelan, J.; Van Breusegem, F. Prohibitins: Mitochondrial Partners in Development and Stress Response. *Trends Plant Sci.* **2010**, *15*, 275–282.

Vierling, E. The Roles of Heat Shock Proteins in Plants. *Annu. Rev. Plant Physiol. Plant Mol. Biol.* **1991**, *42*, 579–620.

Wahid, A.; Gelani, S.; Ashraf, M.; Foolad, M. R. Heat Tolerance in Plants: An Overview. *Environ. Exp. Bot.* **2007**, *61*, 199–223.

Wang, H.-B.; Wong, M.-H.; Lan, C.-Y.; Qin, Y.-R.; Shu, W.-S.; Qiu, R.-L.; Ye, Z.-H. Effect of Arsenic on Flavonoid Contents in *Pteris* Species. *Biochem. Syst. Ecol.* **2010**, *38*, 529–537.

Weisany, W.; Sohrabi, Y.; Heideri, G.; Siosemardeh, A.; Ghassemi-Golezani, K. Changes in Antioxidant Enzymes Activity and Plant Performance by Salinity Stress and Zinc Application in Soybean (*Glycine max* L.). *Plant Omics J.* **2012**, *5*, 60–67.

Werner, T.; Nehnevajovaa, E.; Keollmera, I.; Novakb, O.; Strnadb, M.; Kreamerc, U.; Schmeullinga, T. Root-specific Reduction of Cytokinin Causes Enhanced Root Growth, Drought Tolerance, and Leaf Mineral Enrichment in Arabidopsis and Tobacco. *Plant Cell* **2010**, *22*, 3905–3920.

Wilkinson, S.; Davies, W. J. Drought, Ozone, ABA and Ethylene: New Insights from Cell to Plant to Community. *Plant, Cell Environ.* **2010**, *33*, 510–525.

Xie, Z. J.; Jiang, D.; Cao, W. W.; Dai, T. B.; Jing, Q. Relationships of Endogenous Plant Hormones to Accumulation of Grain Protein and Starch in Winter Wheat Under Different Post-anthesis Soil Water Statuses. *Plant Growth Regul.* **2003,** *41,* 117–127.

Xiong, L.; Wang, R. G.; Mao, G.; Koczan, J. M. Identification of Drought Tolerance Determinants by Genetic Analysis of Root Response to Drought Stress and Abscisic Acid. *Plant Physiol.* **2006,** *142,* 1065–1074.

Xue, G. P.; McIntyre, C. L.; Jenkins, C. L.; Glassop, D.; van Herwaarden, A. F.; Shorter, R. Molecular Dissection of Variation in Carbohydrate Metabolism Related to Water-soluble Carbohydrate Accumulation in Stems of Wheat. *Plant Physiol.* **2008,** *146,* 441–454.

Yamane, Y.; Kashino, Y.; Koike, H.; Satoh, K. Effects of High Temperatures on the Photosynthetic Systems in Spinach, Oxygen-evolving Activities, Fluorescence Characteristics and the Denaturation Process. *Photosynth. Res.* **1998,** *57,* 51–59.

Zarrouk, O.; Francisco, R.; Pinto-Marijuan, M.; Brossa, R.; Santos, R. R.; Pinheiro, C.; Costa, J. M.; Lopes, C.; Chaves, M. M. Impact of Irrigation Regime on Berry Development and Flavonoids Composition in Aragonez (Syn. Tempranillo) Grapevine. *Agric. Water Manage.* **2012,** *114,* 18–29.

Zhang, F.-Q.; Wang, Y.-S.; Lou, Z.-P.; Dong, J.-D. Effect of Heavy Metal Stress on Antioxidative Enzymes and Lipid Peroxidation in Leaves and Roots of Two Mangrove Plant Seedlings (*Kandelia candel* and *Bruguiera gymnorrhiza*). *Chemosphere* **2007,** *67,* 44–50.

Zhang, S. W.; Li, C. H.; Cao, J.; Zhang, Y. C.; Zhang, S. Q.; Xia, Y. F.; Sun, D. Y.; Sun, Y. Altered Architecture And Enhanced Drought Tolerance in Rice via the Down-regulation of Indole-3-acetic Acid by TLD1/OsGH3.13 Activation. *Plant Physiol.* **2009,** *151,* 1889–1901.

Zhou, L.; Li, C.; Zhang, X.; Johnson, R.; Bao, G.; Yao, X.; Chai, Q. Effects of Cold Shocked Epichloe Infected *Festuca sinensis* on Ergot Alkaloid Accumulation. *Fungal Ecol.* **2015,** *14,* 99–104.

Zhu, J.-K. Plant Salt Tolerance. *Trends Plant Sci.* **2001,** *6,* 66–71.

CHAPTER 2

SIGNIFICANCE OF SOIL ORGANIC MATTER IN RELATION TO PLANTS AND THEIR PRODUCTS

SUBHASISH DAS and SATYA SUNDAR BHATTACHARYA*

Department of Environmental Science, Tezpur University, Tezpur 784028, Assam, India

Corresponding author, Tel.: +91 3712 267007/+91 3712 2670078/+91 3712 2670079x5610; E-mail: satyasundarb@yahoo. co.in; satya72@tezu.ernet.in.

CONTENTS

2.1   SOIL ORGANIC MATTER

Soil organic matter (OM) has been used as plant growth promoter and regulator since the advent of human civilization. Considering the history of plant sciences, much prior to the auxin concept establishment, the term "auximones" was coined to describe plant growth-promoting humic acids extracted from peat (Zandonadi et al., 2013). The biggest and original source of soil OM is plant tissue. Under natural conditions, the tops and roots of trees, shrubs and grasses, and other native plants annually supply large quantities of organic residues. Soil organisms decompose all these organic materials which eventually become part of the underlying soil by infiltration or physical incorporation (Lee et al., 2014). Comparatively, animals, however, are considered secondary sources of OM after plants. The animal feces or the waste products are incorporated into the soil system after decay and detritivory. Certain forms of animal life, especially the earthworms, termites, ants, and others, also play an important role in the translocation of soil and plant residues within the soil system (Filley et al., 2008; Dowuona et al., 2012). Figure 2.1 portrays a schematic representation of soil OM. In addition to other soil factors, the OM content of a soil is intimately related to its productivity. Soil OM is an important feature of the soil system which varies significantly with soil type, climatic conditions, and land-use pattern (Craswell & Lefroy, 2001). The soil OM supports the soil system in various ways. Some of the functional significances of soil OM are as follows (Brady, 1995):

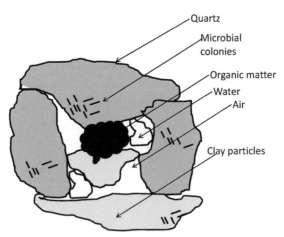

FIGURE 2.1   Schematic representation of soil organic matter complex.

- OM acts as a storehouse of plant nutrients (N, P, S, etc.);
- OM increases exchange capacity of cations and anions;
- OM provides energy for microbial activity;
- OM increases water-holding capacity of soil;
- OM improves soil structure;
- OM reduces crusting and increases infiltration;
- OM reduces the effects of compaction; and
- OM buffers the soil against rapid changes in acidity, alkalinity, and salinity.

The above stated factors coordinately impart "tilth" to the soil. The production of large quantities of residues, and their subsequent decay, is necessary for good soil tilth and productivity.

2.2 DECOMPOSITION OF OM

The decomposition or disintegration of OM is a long-time process. Various biotic and abiotic factors influence the process of OM decomposition. Climatic factors like temperature and moisture play an important role in the decomposition process (Conant et al., 2011; Gabriel & Kellman, 2014). Soil clay content supports a faster rate of OM decomposition in soil (Wei et al., 2014). Apart from the stated factors, many soil organisms also modify the soil system by full or partial decomposition of OM (Xu et al., 2015). This is due to the complex architecture of the various components of OM. The general composition of soil OM involves carbohydrates (starch, cellulose, hemi-cellulose, etc.), lignin (found in woody plant tissue), fats and oils (found primarily in seeds), crude protein, crude fiber, and others. Organic compounds vary greatly in their rate of decomposition. They may be listed in terms of ease of decomposition as follows (Brady, 1995):

- Sugars, starches, and simple proteins   rapid decomposition
- Crude proteins
- Hemicelluloses
- Cellulose
- Fats, waxes, etc.
- Lignins   very slow decomposition

The initiation of decomposition process of organic materials is triggered with the addition of fresh plant tissue to soil. The sugars and simple proteins decompose most readily; at the other extreme; lignins are the most resistant to breakdown. When organic tissue is added to soil, three general reactions take place (Brady, 1995):

- The biggest proportion of material faces enzymatic oxidation with the liberation of CO_2, H_2O, energy, and heat.
- Mineralization or immobilization of macro or microelements occurs through element-specific reactions.
- Reconstitution of complex organic compounds takes place with the involvement of plant and microbial entities and these compounds collectively comprise soil humus.

Decomposition of OM is an oxidation process when all the organic compounds found in residues are disintegrated with evolution of energy. The oxidation reaction can be summed as follows.

$$\underset{\text{C,H-containing compounds}}{-\left(C,4H\right)} + 2O_2 \xrightarrow{\text{enzymatic oxidation}} CO_2 + 2H_2O + \text{Energy}$$

Many intermediate steps are involved in this overall reaction, and it is accompanied by important side reactions that involve elements other than C and H. This basic reaction answers the complex decomposition pathway of OM in the soil as well as the oxygen consumption and release of CO_2. For example, during protein decay amino acids like glycine and cysteine are produced apart from CO_2 and H_2O. In turn, these amino acids are further broken down by microbial action, eventually yielding simple inorganic ions like NH_4^+, NO_3^-, and SO_4^{2-} (Wheeler, 1963).

2.3 CARBON CYCLE

As OM decays, CO_2 is among the immediate breakdown products. C is a common constituent of all OM and is involved in essentially all life processes. Thus, the transformation of this element is also termed as "biocycle" that makes possible the continuity of life on earth (Fig. 2.2). The soil OM influences various soil processes like evolution and consumption of greenhouse gases (CH_4, N_2O, etc.) (Horwath, 2015). Carbon cycling in soil is under strict microbial operation (Schimel & Schaeffer, 2012). There are several components in the C-cycle machinery termed as pool, stock or reservoir, and flux. The pool is the store house for large amount of C operating in the earth system. Soil is a large pool for C sequestration, holding about 75% of the C operating in the atmosphere. The tropical forests represent one-third of the total metabolic activity on the land surface (Malhi, 2012). Any movement of C between the reservoirs is called flux. In any integrated system, fluxes connect reservoirs together to create cycles and feedbacks.

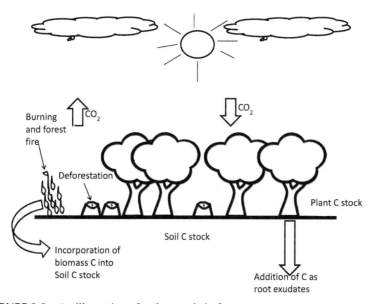

FIGURE 2.2   An illustration of carbon cycle in forest ecosystem.

2.3.1   RELEASE OF CO_2

Through the process of photosynthesis, CO_2 is stored and incorporated by higher plants and resynthesized into numerous organic compounds (Simpraga et al., 2011). When these organic compounds reach the soil with plant residues, they undergo digestion, and CO_2 is liberated. Microbial activity is the main soil source of CO_2, although appreciable amounts come from respiration of plant roots, and small amounts are brought down dissolved in rain water (Rastogi et al., 2002). Much of the CO_2 of the soil ultimately escapes to the atmosphere, where it can again be used by plants, thus completing the cycle. A meager amount of CO_2 reacts in the soil that produces carbonic acid, carbonates, and bicarbonates of Ca, K, Mg, and other base-forming cations. The bicarbonates are readily soluble and can be removed in drainage or used by higher plants (Tyler & Olsson, 2001).

2.3.2 OTHER CARBON PRODUCTS OF DECAY

The decay of OM forms a series of carbon products. Small quantities of elemental C are found in soil (Nicolas et al., 2012). Under highly anaerobic conditions, CH_4 and CS_2 may be produced in small amounts. But of all the simple C products, CO_2 is by far the most abundant. The C-cycle is all inclusive because it involves soil and its flora and fauna.

2.4 SOIL HUMUS

Humus is a complex and rather resistant mixture of brown amorphous and colloidal organic substances that results from microbial decomposition and synthesis and has chemical and physical properties of great significance to soils and plants (Brady, 1995). There are different kinds of humus forms found in the soils around the globe. The most popular forms of humus are mull, moder, and mor (Ponge, 2013). Mixing of the soil OM, minerals, plant exudation, animals, and microbial excreta lead to the formation of a special type of humus in the upper crumby soil layer called "mull." Again, when soil OM only consists of fungal mycelia and animal excreta and lies over the mineral layer of soil, it is referred to as "moder." When transformation of plant litter occurs slowly leading to its accumulation in the "E" horizon (purely mineral horizon) or in the parent rock, it is called "mor" (Fig. 2.3). Some other forms of humus are "amphi" and "tangel" which are mainly described on calcareous parent rocks in Mediterranean and sub-alpine climatic conditions (Galvan et al., 2008).

The complex compounds that make up humus are not merely degraded plant materials but mostly the bulk of these compounds have resulted from two general types of biochemical reactions: decomposition and synthesis. Decomposition starts as chemicals in the plant residues which are further subjected to disintegration by soil organisms. Even lignin is degraded by the intense microbial reactions. The second humus-forming step is termed as biochemical synthesis when the simpler organic compounds coming from the breakdown of the parent organic complexes get involved. These simpler chemicals are metabolized into new compounds in the body tissue of the soil microorganisms (Matsumura, 1987). The new compounds are subject to further modification and synthesis as the microbial tissue is subsequently attacked by other soil microorganisms. Some other additional synthetic reactions involve such breakdown products of lignin as the phenols and quinines (Camarero et al., 1994). These decomposition products, present initially as separate molecules called monomers, are enzymatically stimulated to join together into polymers. By this process of polymerization, polyphenols and polyquinones are formed (Prusty & Azeez, 2015). These high molecular weight compounds interact with N-containing amino compounds and give rise to a significant component of resistant humus. The formation of these polymers is encouraged by the presence of colloidal clays.

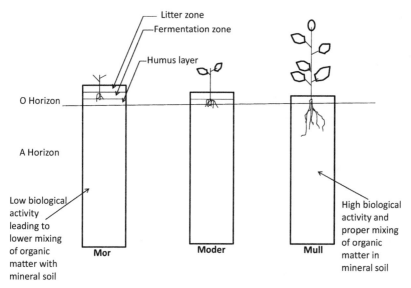

FIGURE 2.3   Schematic diagram of three major types of humus in soil

The humic substances make up to 60–80% of the soil OM (Trevisan et al., 2010). Humic substances are characterized by aromatic structures and include polyphenols and polyquinones, which are even more complex. The humic substances have no sharply defined physical or chemical properties, which are characteristic of more simple nonhumic compounds (Prusty & Azeez, 2015). They are amorphous, dark in color, and have very high molecular weights. On the basis of resistance to degradation and of solubility in acids and alkalis, humic substances have been classified into three chemical groupings: (a) fulvic acid, (b) humic acid, and (c) humin. Fulvic acid has the lowest molecular weight of the three, is soluble in both acid and alkali, and is most susceptible to microbial attack. Humic acid is medium in molecular weight and color, soluble in alkali, insoluble in acid, and intermediate in resistance to microbial degradation. Humin has the highest molecular weight, darkest in color, insoluble in both acid, and alkali, and most resistant to microbial attack (Brady, 1995). Despite differences in chemical and physical makeup, the three humic groups have some similarities with regard to properties, such as ability to absorb and release cations, and hence they are collectively called humic substances.

Apart from the humic group, the soil OM consists of a considerable proportion of nonhumic substances. The nonhumic are less complex and less resistant to microbial attack compared to humic substances. They are comprised of specific organic compounds with definite physical and chemical properties. Some of these nonhumic materials are only modified by microbial action while others are synthesized as break-down occurrs. The usual nonhumic compounds in soil are polysaccharides, polymers that have sugar-like structures, polyuronides, and so on. Some simpler compounds like organic acids and protein-like materials are also part of nonhumic substances. The humic and the nonhumic components of the OM continually interact with each other. Reactions occur incessantly in the soil system. During the formation of humus, all the intermediate structural compounds continuously react with each other whether humic or nonhumic in nature. For example, proteins and other N-compounds react with a wide variety of organic compounds in both groups, including the humic acids and polysaccharides. These reactions are essential for N-conservation in soils because in the resultant compounds the protein N is somehow protected from microbial attack.

Humic acid is an important soil component that bestows fertility to it and leads to better plant growth and productivity (Dhanpal & Sekar, 2013). Interaction with some layer silicate clays is another way of stabilizing soil N. These clays are known to attract and hold substances such as amino acids, peptides, and proteins, forming complexes that protect the N-containing compounds from microbial decomposition. Specific protein–clay linkages are not the only important reactions that occur between organic and inorganic compounds in soils. In some cases, clays seem to catalyze some oxidation and polymerization reactions even in the absence of microorganisms (Lee et al., 2004). Also various organomineral complexes involve linkage of phenols or organic acids with silicate clays or iron and aluminum oxides. Humus is subject to continual microbial attack. Without the annual addition of organic residues, microbial action results in a reduction in soil OM levels (Ma et al., 2012). Each time, newly synthesized polymers join those formed in previous time periods. Due to its resistivity toward easy decay, humic compounds are important in maintaining OM levels in soil and also help protect N and other essential elements that are bound in the humic complex.

2.4.1 HUMUS–COLLOIDAL CHARACTERISTICS

The humus–colloids are stable soil aggregates which are formed by the mating of mineral nutrients and humic substances. Such complexes are formed by microbial action. Soil pH has an important role in the complexation of humus–colloid (Varadachari et al., 1994). The major characteristics of the humus–colloidal constituents of soil are as follows (Brady, 1995):

- The tiny colloidal humus particles are composed of C, H, and O mostly in the forms of polyphenols, polyquinones, polyuronides, and polysaccharides.
- The surface area of humus colloids per unit mass is very high, generally exceeding that of silicate clays.
- The colloidal surfaces of humus are negatively charged, the sources of the charge being hydroxyl, carboxylic, or phenolic groups. The extent of the negative charge is pH dependent.
- At high pH, the cation exchange capacity of humus on a mass basis far exceeds that of most silicate clays.
- The water-holding capacity of humus on a mass basis is greater than silicate clays.
- Humus has a very favorable effect on aggregate formation and stability.
- The dark brown color of humus tends to distinguish it from most of the other constituents in soil.

- Cation exchange reactions with humus are qualitatively similar to those occurring with silicate clays.

The humic micelles, like clay particles, carry a swarm of adsorbed cations. Thus, colloidal humus may be represented by the same illustrative formula used for clay.

Humus enhances mineral breakdown and, in turn, nutrient availability in two ways. First, humic acids can attack the minerals and bring about their decomposition, thereby releasing essential base forming cations.

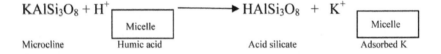

$$KAlSi_3O_8 + H^+ \xrightarrow{\text{Micelle}} HAlSi_3O_8 + K^+ \quad \text{Micelle}$$

Microcline   Humic acid   Acid silicate   Adsorbed K

The K is changed from a molecular to an adsorbed state, which is more readily available to higher plants. The second mechanism for increasing the availability of some cations is through the formation of stable organomineral complexes with these ions. For example, polysaccharides and fulvic acids form such complexes with metallic ions such as Fe^{3+}, Cu^{2+}, Zn^{2+}, and Mn^{2+}. The cations are attracted from the minerals in which they are found and held in complex form by the organic molecule. Later, they may be taken up by plants or may take part in the synthesis of clay and other inorganic constituents (Pokrovsky & Schott, 2002).

2.5 INFLUENCE OF ORGANIC COMPOUNDS ON PLANT GROWTH AND METABOLITE PRODUCTION

Important aspects of soil–plant interactions involve the regulating processes that control the accumulation or disintegration of OM coupled with the genesis or disappearance of soil horizons. OM is highly plant available. For example, certain amino acids like alanine, glycine, and so on, can be directly taken up by the plant roots for usage. The concentration of these amino acids differs highly from soil to soil. In normal conditions, concentration of such amino acids is very meager in amount to meet the plant demand for N. Organic acids like vanillic acids, carboxylic acids, and others, are highly significant for plant use (Brady, 1995). Organic compounds in soil, though present in little amount contain various growth-promoting substances. Humic substances also influence the expression of various plant metabolites. Merlo et al. (1991) reported the positive influence of soil humic substances on the elevated carbohydrate synthesis in maize foliage. A schematic representation of some important plant secondary metabolites is provided in Figure 2.4.

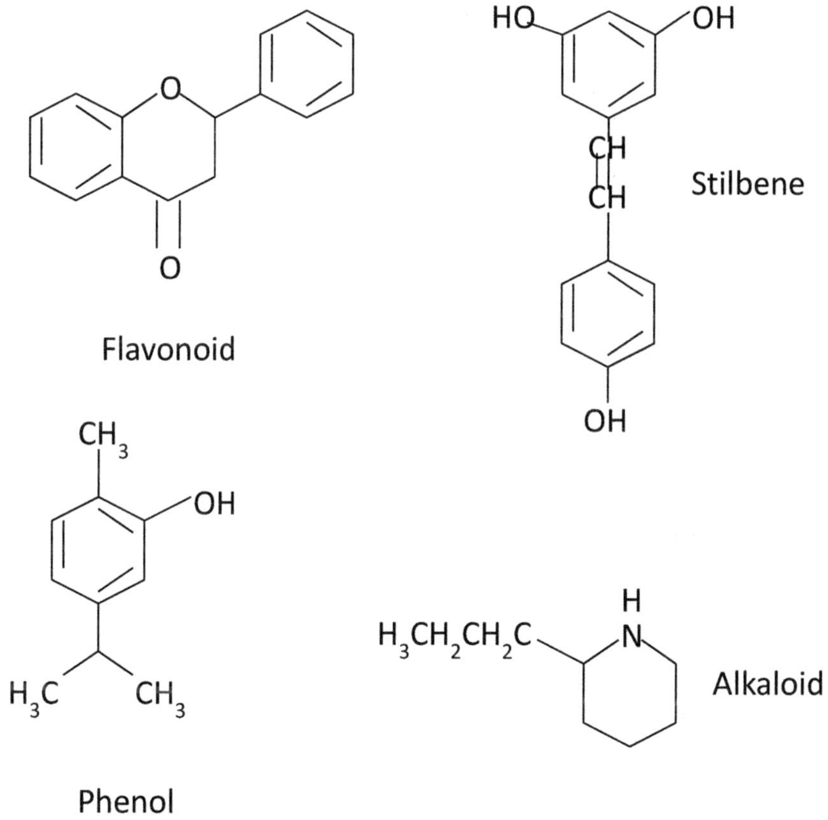

FIGURE 2.4 Schematic representation of some typical plant secondary metabolites.

Organic fertilization imparts high OM in soil which in turn influences the higher production of plant phenols, flavonoids, stilbenes, and so on (Ibrahim et al., 2013). Similar research by Ali et al. (2012) highlights the positive impact of OM on the esculin (bioactive compound) content in *Cichorium intybus* plants. Application of humic substances as foliar spray in tomato enhanced nitrate uptake along with the nitrate reductase activity coupled with the activation of secondary metabolic phenylalanine ammonia lyase pathway (Olivares et al., 2015). Edreva et al. (2008) reported the chemical characteristics and functions of certain plant metabolites (provided in Table 2.1). Application of OM in soil also enhances the flavonolignan (plant metabolite) production in milk thistle (Afshar at al., 2014). Application of agrowaste in the soil system elicits the synthesis of various amino acids and other important metabolites in plant body (Ibrahim & Mumtaz, 2014). Soil humic substances also impart higher plant defense toward pathogens (Berbara & Garcia, 2014). Incorporation of humic substances along with certain bacterial strains results in elevation of ATPase enzyme activity in plants, better N metabolism, and greater net photosynthesis (Canellas et al., 2012). OM decomposition is a time-dependent process. After decomposition, different classes of growth-promoting compounds like vitamins, amino acids, plant hormones (auxins, gibberellins, etc.) are liberated into the soil system. With the surge of such nutritious substances, plant and microbial growth in soil is enhanced manifold. Incorporation of vermistabilized or composted industrial waste products holds much good in boosting plant growth and metabolism. Recent reports show that toxic wastes like paper mill waste, olive mill waste, waste-water sludge, distillery sludge, municipal solid waste, and so on, after vermitreated influence better productivity in plants (Sahariah, et al., 2014; Garcia-Gomez et al., 2002; Vivas et al., 2009; Singh & Suthar, 2012; Ramana et al., 2002). A gist of the influence of various organic inputs on plant metabolite synthesis is provided in Table 2.2.

Plant metabolites like flavonoids, phenols, tannins, stilbenes, alkaloids, and so on, and various enzymes are significantly influenced by the high nutrient content of the OM amendments applied to the soil or other growing media (Naguib et al., 2012; Taie et al., 2008). Table 2.3 shows the effect of different types of vermicompost on plant growth and metabolite production. The antioxidant concentration in crops is also enhanced by the application of organic manure in spite of regular chemical fertilizers (Bimova & Pokluda, 2009). Bilalis et al. (2009) reported that nicotine production in *Nicotiana tabaccum* is stimulated by the incorporation of organic manure and better irrigation practices. Incorporation of organic manure in soil elevates plant defense capacity against pathogens along with high production of metabolites (Bommesha et al., 2012).

TABLE 2.1   Different Structural Forms of Bio-active Compounds in Plants.

Sr. No.	Chemical characteristics	Functions
1	Presence of: -COOH, -OH, -NH$_2$ 	Structure stabilizing, anti-microbial, hypersensitivity providing
2	Availability of Ring closed structures CH=CH=C=O, OH Short side chains Long side chains -CH=CH=CH=CH-	Photo-protective
3	Availability of : -NH$_2$, -SH, -OH -CH-CH=CH-CH- Unsaturated carbon chains 	Antioxidant, antiradical

TABLE 2.2   Effect of Different Organic Fertilizers on Plant Metabolites.

Sr. No.	Organic fertilizers	Plant metabolites	Reference
1	Farm yard manure	Carotenoid and ascorbic acid content increased	Kipkosgei et al. (2003)
2	Green manure	High chlorophyll and fatty acid content	Mohammadi et al. (2011)
3	Vermicompost	Essential oil content increased	Singh et al. (2013)
4	Poultry manure	Increased protein, lipid, and carbohydrate concentration	Oyedeji et al. (2014)
5	Compost	High chlorophyll and leaf nutrient content	Belda et al. (2013)
6	Biochar	Increased content of tannins and organic acids	Schmidt et al. (2014)
7	Swine manure	High chlorophyll and leaf nutrient content	Kanto et al. (2012)
8	Compost tea	High sugar and phytonutrient concentration	Pant et al. (2012)
9	Tree bark extract	High pigment content in plants	Tanase et al. (2014)
10	Sea weed	High N assimilation and basal metabolism	Gonzalez et al. (2013)
11	Humic acid	Increased chlorophyll content	Zhang et al. (2014)
12	Pruning waste	Increased chlorophyll content	Morales-Corts et al. (2014)

TABLE 2.3   Effect of Vermicompost on Expression of Different Plant Metabolites.

Sr. No.	Type of vermicompost	Used plants	Metabolite	Reference
1	Cattle manure vermicompost	Pepper mint	Increased phenol, oil, chlorophyll, and carotenoid content	Ayyobi et al. (2013)
2	Commercial vermicompost	Bitter gourd	Increased phenol and flavonoid content	Benitez et al. (2013)
3	Pig manure vermicompost	Tomato and marigold	Enhanced chlorophyll content	Atiyeh et al. (2000)
4	Vermiwash	Chilly	High phytonutrients	Khan et al. (2014)
5	Vermicompost tea	Tomato	Enhanced nitrate reductase activity	Márquez-Quiroz et al. (2008)

Along with its beneficial effects, organic substances may also lead to the formation of some toxic compounds (Toussoun et al., 1967). For example, dihydroxystearic acid that is produced as a by-product during OM degradation in soil is highly toxic to higher plants. However, with the incorporation of various soil management practices like efficient drainage, tillage, application of lime and fertilizers, the problem of soil toxicity can be nullified to a considerable extent. The major influences of soil OM on the soil system can be studied under the following heads:

- Application of OM to soil renders the soil color black from brown,
- OM enhances soil physical conditions like granulation and water retention capacity coupled with the reduction of cohesion and plasticity,
- Betterment of soil cation exchange capacity is evident after applying OM to the soil system,
- The capacity of soil to supply and replenish the nutrient deficiency situations is increased by the incorporation of OM. OM not only acts as a source of nutrients for plants but also aids in the weathering of minerals through involvement of various organic acids.

Thus, OM can be regarded as a "store house" of nutrients for plants. Application of required OM coupled with the congenial environmental conditions may render better soil health and luxuriant plant growth.

2.6   FACTORS AFFECTING SOIL ORGANIC MATTER

Soil OM and soil N are affected by a series of important factors. Primarily, the soil OM is influenced by the type and quantity of plant or animal organic residues added to the soil system (Baldock, 2012). Some of the major factors which affect the soil OM are summarized as under:

- Climate
- Vegetation
- Soil physical properties
- Cropping and tillage
- Crop rotation, residues and plant nutrients
- Stubble burning or removal
- Overgrazing

2.6.1   INFLUENCE OF CLIMATE

Climate influences the dynamics of OM and its fractions in the soil system (Franzluebbers et al., 2001). Major climatic factors like temperature and precipitation play a prominent role on the levels of N and OM content in soil. If one moves from warmer to cooler regions, the amount of unmineralized OM and N will increase. This is because the warmer humid regions are more rich in microbes that lead to quick decomposition of the organic substrates leading to a lower C/N ratio than the cold regions where microbial activity is very low. In general, under comparable soil moisture levels the soil OM and N increase by 2–3 times by every 10°C decline in temperature (Buchan, 2014). In case of highly cultivated and well-drained soil, decomposition rate is high, accelerating the nutrient release but low residual C content. Like temperature, soil water content also holds huge influence on the OM and N accumulation capacity of soils. Clearly, with increasing soil moisture the amount of OM and N increase in soil. Thus, the synergistic effect of temperature and rainfall exert positive influence on the accumulation of OM and N in soil.

2.6.2   INFLUENCE OF VEGETATION

Natural vegetation overlying a soil type affects it positively with respect to its OM and N content. Vegetation types like grassland, forest, and so on, exert differential impact on the soil OM and N content (Pulido-Fernandez et al., 2013). Comparatively, soils under grasses accumulate more N and OM than forests. The sparse vegetation type in grasslands allows significantly lesser rate of decomposition of organic complexes promoting a high soil OM and N compared to the forest soil.

2.6.3   EFFECT OF SOIL PHYSICAL PROPERTIES

Soil physical properties like texture, drainage, aeration, and others, play a crucial role on the accumulation of OM and N in soil (Buschiazzo et al., 2004). Soil OM and N are comparatively high in fine-textured soil than coarse-textured soil. Thus, soil with more clay and silt percentage offer more OM accumulation potential than those with sand- and gravel-rich soil. The inherent interaction between clay and organomineral complexes is high which safeguards the organic N in the form of clay-bound protein complex. This results in enhanced OM contents. Moreover, soils with good drainage and aeration are poor in OM content.

2.6.4   EFFECT OF CROPPING AND TILLAGE

Soils that undergo land-use change portray significant alteration in the level of OM and N (Curaqueo et al., 2010). Undisturbed natural soil is rich in OM than soil under cultivation. Virgin soil is high in OM and N because all the organic input from the overlying natural vegetation is straightly returned to the soil. On the other hand, soil under intensive agriculture, breaks the trend of OM and N accumulation due to high microbial activity and associated decomposition. Both conventional and modern tillage practices impart a variable effect on the soil OM content. The modern tillage practices lower the erosion risk and discourage the rapid decay of crop residues which maintains the OM budget of the soil for a longer time.

2.6.5   EFFECT OF CROP ROTATION, RESIDUES, AND PLANT NUTRIENTS

Sequence of cropping, incorporation of fertilizers, and other organic amendments influence the accumulation capacity of OM and N in soil (Campbell et al., 1991; Bierke et al., 2008). Soil that are well managed for years with the addition of appropriate amount of fertilizers, lime, and OM stay productive for long time and thus maintain a sound soil OM and N budget. On the other hand, the less productive or unproductive soils are poor in OM content. The extent of soil productivity depends highly on the amount of root and top soil residues that are returned to the soil after decay. Thus, the most productive arable soil will be low in OM compared to an undisturbed natural soil patch.

2.6.6   EFFECT OF STUBBLE BURNING OR REMOVAL

Stubble burning is a fire-setting pattern in agricultural fields after harvesting a crop. There are two main consequences of stubble burning in soil system. First, most of the plant carbon is lost in burning as CO_2. Second, the black carbon formed after burning is added to the soil. Burning of crop residues leads to minimal increase in soil organic carbon content along with high earthworm population (Virto et al., 2007).

2.6.7 EFFECT OF OVERGRAZING

Overgrazing is a situation that arises when excessive amount of plant individuals are removed by primary consumers leading to desertification. Such incidences are of usual occurrence mainly in the semi-arid regions where vegetation is very sparse. Overgrazing directly influences the microbial communities in soil, thus hampering important soil processes like decomposition, mineralization, and others (Abril & Bucher, 1999). Thus, OM accumulation is affected by overgrazing. Overgrazing also leads to soil compaction, low water retention capacity, aeration, and so on.

KEYWORDS

- soil organic matter
- disintegration
- climate
- natural vegetation
- overgrazing

REFERENCES

Abril, A.; Bucher, E. H. The Effects of Overgrazing on Soil Microbial Community and Fertility in the Chaco Dry Savannas of Argentina. *Appl. Soil Ecol.* **1999,** *12,* 159–167.

Afshar, R. K.; Chaichi, M. R.; Assareh, M. H.; Hashemi, M.; Liaghat, A. Interactive Effect of Deficit Irrigation and Soil Organic Amendments Onseed Yield and Flavonolignan Production of Milk Thistle (*Silybum marianum* L. Gaertn.). *Ind. Crop Prod.* **2014,** *58,* 166–172.

Ali, Z.; Ganie, S. H.; Narula, A.; Abdin, M. Z.; Srivastava, P. M. Organic and Inorganic Fertilizers Influence Biomass Production and Esculin Content in *Cichorium intybus* L. *J. Phytol.* **2012,** *4,* 55–60.

Atiyeh, R. M.; Subler, S.; Edwards, C. A.; Bachman, G.; Metzger, J. D.; Shuster, W. Effects of Vermicomposts and Composts on Plant Growth in Horticultural Container Media and Soil. *Pedobiologia* **2000**, *44*, 579–590.

Ayyobi, H.; Peyvast, G.–A.; Olfati, J. A. Effect of Vermicompost and Vermicompost Extract on Oil Yield and Quality of Peppermint (*Mentha piperita* L.). *J. Agri. Sci.* **2013**, *58*, 51–60.

Baldock, J. A.; Wheeler, I; McKenzie, N.; McBrateny, A. Soils and Climate Change: Potential Impacts on Carbon Stocks and Greenhouse Gas Emissions, and Future Research for Australian Agriculture. *Crop Past. Sci.* **2012**, *63*, 269–283.

Belda, R. M.; Mendoza-Hernández, D.; Fornes, F. Nutrient-Rich Compost versus Nutrient-Poor Vermicompost as Growth Media for Ornamental-Plant Production. *J. Plant Nutr. Soil Sci.* **2013**, *176*, 827–835.

Benitez, M. M.; Zara, R. R.; de Guzman, C. C. Comparative Effects of Soil Organic Amendments on Growth, Yield and Antioxidant Content of Bitter Gourd (*Momordica charantia* L. cv. Makiling). *Phillip. Agric. Sci.* **2013**, *96*, 359–369.

Berbara, R. L. L.; Garcia, A. C. Humic Substances and Plant Defense Metabolism. In *Physiological Mechanisms and Adaption Strategies in Plants under Changing Environment*; Ahmad, P., Wani, M. R., Eds.; Springer: New York, 2014; pp 297–319.

Bierke, A.; Kaiser, K.; Guggenberger, G. Crop Residue Management Effects on Organic Matter in Paddy Soils—The Lignin Component. *Geoderma* **2008**, *146*, 48–57.

Bilalis, D.; Karkanis, A.; Efthimiadou, A.; Konstantas, A.; Triantafyllidis, V. Effects of Irrigation System and Green Manure on Yield and Nicotine Content of Virginia (flue-cured) Organic Tobacco (*Nicotiana tabaccum*), Under Mediterranean Conditions. *Ind. Crops Prod.* **2009**, *29*(2), 388–394.

Bimova, P.; Pokluda, R. Impact of Organic Fertilizers on Total Antioxidant Capacity in Head Cabbage. *Hortic. Sci.* **2009**, *36*, 21–25.

Bommesha, B.; Naik, M. I.; Mutthuraju, G. P.; Pannure, A.; Imran, S.; Prashantha, C. Effect of Organic Manures on Biochemical Components of Pigeon Pea, *Cajanus cajan* (L.) Millsp. and Their Impact on the Incidence of Insect Pests. *Curr. Biot.* **2012**, *6*, 171–180.

Brady, N. C. *The Nature and Properties of Soils*, 10th ed.; Prentice Hall of India: New Delhi, 1995.

Buchan, G. D. Temperature Effects in Soil. *Encyclopedia of Agrophysics*; Springer: Netherlands, 2014; pp 891–895.

Buschiazzo, D. E.; Estelrich, H. D.; Aimar, S. B.; Viglizzo, E.; Babinec, F. J. Soil Texture and Tree Coverage Influence on Organic Matter. *J. Range Mange.* **2004**, *57*, 511–516.

Camarero, S.; Galletti, G. C.; Martinez, A. T. Preferential Degradation of Phenolic Lignin Units by Two White-Rot Fungi. *Appl. Environ. Microbiol.* **1994**, *60*, 4509–4516.

Campbell, C. A.; Biederbeck, V. O.; Zentner, R. P.; Lafond, G. P. Effect of Crop Rotations and Cultural Practices on Soil Organic Matter, Microbial Biomass and Respiration in a Thin Black Chernozem. *Can. J. Soil Sci.* **1991**, *71*, 363–376.

Canellas, L. P.; Balmori, D. M.; Medici, L. O.; Aguiar, N. O.; Campostrini, E.; Rosa, R. C. C.; Facanha, A. R.; Olivares, F. L. A Combination of Humic Substances and *Herbaspirillum seropedicae* Inoculation Enhances the Growth of Maize (*Zea mays* L.). *Plant Soil* **2012**. http://dx.doi.org/10.1007/s11104-012-1382-5.

Conant, R. T.; Ryan, M. G.; Agren, G. I.; Birge, H. E.; Davidson, E. A.; Eliasson, P. E.; Evans, S. E.; Frey, S. D.; Giardina, C. P.; Hopkins, F. M.; Hyvonen, R.; Kirschbaum,

M. U. F.; Lavallee, J. M.; Leifeld, J.; Parton, W. J.; Steinweg, J. M.; Wallenstein, M. D.; Wetterstedt, J. A. M.; Bradford, M. A. Temperature and Soil Organic Matter Decomposition Rates—Synthesis of Current Knowledge and a Way Forward. *Global Change Biol.* **2011,** *17,* 3392–3404.

Craswell, E. T.; Lefroy, R. D. B. The Role and Function of Organic Matter in Tropical Soils. *Nutr. Cycl. Agroecosyst.* **2001,** *61,* 7–18.

Cruz-Lázaro, E.; Reyes-Carrillo, J. L. Effect of Vermicompost Tea on Yield and Nitrate Reductase Enzyme Activity in Saladette Tomato. *J. Soil Sci. Plant Nutr.* **2014,** *14,* 223–231.

Curaqueo, G.; Acevedo, E.; Cornejo, P.; Seguel, A.; Rubio, R.; Borie, F. Tillage Effect on Soil Organic Matter, Mycorrhizal Hyphae and Aggregates in a Mediterranean Agroecosystem. *R. C. Suelo Nutr. Veg.* **2010,** *10,* 12–21.

Dhanpal, S.; Sekar, D. S. Humic Acids and Its Role in Plant Tissue Culture at Low Nutrient Level. *J. Acad. Indus. Res.* **2013,** *2,* 338–340.

Dowuona, G. N. N.; Atwere, P.; Dubbin, W.; Nude, P. M.; Mutala, B. E.; Nartey, E. K.; Heck, R. J. Characteristics of Termite Mounds and Associated Acrisols in the Coastal Savanna Zone of Ghana and Impact on Hydraulic Conductivity. *Nat. Sci.* **2012,** *7,* 423–437.

Edreva, A.; Velikova, V.; Tsonev, T.; Dagnon, S.; Gurel, A.; Aktas, L.; Gesheva, E. Stress-Protective Role of Secondary Metabolites: Diversity of Functions and Mechanisms. *Gen. Appl. Plant Physiol.* **2008,** *34,* 67–78.

Filley, T. R.; McCormick, M. K.; Crow, S. E.; Szlavecz, K.; Whigham, D. F.; Johnston, C. T.; van den Heuvel, R. N. Comparison of the Chemical Alteration Trajectory of *Liriodendron tulipifera* L. Leaf Litter among Forests with Different Earthworm Abundance. *J. Geophys. Res.* **2008.** http://dx.doi.org/10.1029/2007JG000542.

Franzluebbers, A. J.; Haney, R. L.; Honeycutt, C. W.; Arshad, M. A.; Schomberg, H. H.; Hons, F. M. Climatic Influences on Active Fractions of Soil Organic Matter. *Soil Biol. Biochem.* **2001,** *33,* 1103–1111.

Gabriel, C.–E.; Kellman, L. Investigating the Role of Moisture as an Environmental Constraint in the Decomposition of Shallow and Deep Mineral Soil Organic Matter of a Temperate Coniferous Soil. *Soil Biol. Biochem.* **2014,** *68,* 373–384.

Galvan, P.; Ponge, J. E.; Chersich, S.; Zanella, A. Humus Components and Soil Biogenic Structures in Norway Spruce Ecosystems. *Soil Soc. Am. J.* **2008,** *72,* 548–557.

Garcia-Gomez, A.; Bernal, M. P.; Roig, A. Growth of Ornamental Plants in Two Composts Prepared from Agroindustrial Wastes. *Bioresour. Technol.* **2002,** *83,* 81–87.

Gonzalez, A.; Castro,J.; Vera, J.; Moenne, A. Seaweed Oligosaccharides Stimulate Plant Growth by Enhancing Carbon and Nitrogen Assimilation, Basal Metabolism, and Cell Division. *J. Plant Growth Regul.* **2013,** *32,* 443–448.

Horwath, W. Carbon Cycling: The Dynamics and Formation of Organic Matter. In *Soil Microbiology, Ecology, and Biochemistry*; Paul, E. A., Ed.; Academic Press: London, 2015; pp 339–282.

Ibrahim, M. H.; Jaafar, H. Z. E.; Karimi, E.; Ghasemzadeh, A. Impact of Organic and Inorganic Fertilizers Application on the Phytochemical and Antioxidant Activity of Kacip Fatimah (*Labisia pumila* Benth). *Molecules* **2013,** *18,* 10973–10988.

Ibrahim, S.; Mumtaz, E. Application of Agro-waste Products as Organic and Value Added Biofertilizer for Improving Plant Growth. *JPCS.* **2014,** *8,* 35–41.

Kanto, U.; Jutamanee, K.; Osotsapar, Y.; Jattupornpong, S. Effect of Swine Manure Extract on Leaf Nitrogen Concentration, Chlorophyll Content, Total Potassium in Plant Parts and Starch Content in Fresh Tuber Yield of Cassava. *J. Plant Nutr.* **2012**, *35*, 688–703.

Khan, M. H.; Meghvansi, M. K.; Gupta, R.; Veer, V.; Singh, L.; Kalita, M. C. Foliar Spray with Vermiwash Modifies the Arbuscular Mycorrhizal Dependency and Nutrient Stoichiometry ofBhut Jolokia (*Capsicum assamicum*). *PLoS ONE* **2014**, http://dx.doi.org/10.1371/journal.pone.0092318.

Kipkosgei, L. K.; Akundabweni, L. S. M.; Hutchinson, M. J. The Effect of Farmyard Manure and Nitrogen Fertilizer on Vegetative Growth, Leaf Yield and Quality Attributes of *Solanum villosum* (Black nightshade) in Keiyo District, Rift Valley. *Afr. Crop Sci. Proc.* **2003**, *6*, 514–518.

Lee, H.; Fitzgerald, J.; Hewins, D. B.; McCulley, R. A.; Archer, S. R.; Rahn, T.; Throop, H. L. Soil Moisture and Soil-litter Mixing Effects on Surface Litter Decomposition: A Controlled Environment Assessment. *Soil Biol. Biochem.* **2014**, *72*, 123–132.

Lee, J.–F.; Liao, P.–M.; Lee, C.–K.; Chao, H.–P.; Peng, C.–L.; Chiou, C.–T. Clay-Catalyzed Reactions of Coagulant Polymers During Water Chlorination. *J. Colloid Interface Sci.* **2004**, *270*, 381–387.

Ma, L.; Huang, W.; Guo, C.; Wang, R.; Xiao, C. Soil Microbial Properties and Plant Growth Responses to Carbon and Water Addition in a Temperate Steppe: The Importance of Nutrient Availability. *PLoS ONE* **2012**. http://dx.doi.org/10.1371/journal.pone.0035165.

Malhi, Y. The Productivity, Metabolism and Carbon Cycle of Tropical Forest Vegetation. *J. Ecol.* **2012**, *100*, 65–75.

Márquez-Quiroz, C.; López-Espinosa, S. T.; Sánchez-Chávez, E.; García-Bañuelos, M. L.; De la Martinez-Mena, M.; Lopez, J.; Almagro, M.; Boix-Fayos, C.; Albaladejo, J. Effect of Water Erosion and Cultivation on the Soil Carbon Stock in a Semiarid Area of South-East Spain. *Soil Tillage Res.* **2008**, *99*, 119–129.

Matsumura, F. Comparative Metabolism of Mixtures of Chemicals by Animals, Plants, and Microorganisms and Their Significance in the Alteration of Pollutants in the Environment. In *Methods for Assessing the Effects of Mixtures of Chemicals*; Vouk, V. B., Butler, G. C., Upton, A. C., Parke, D. V., Asher, S. C., Eds.; John Wiley & Sons: New York, 1987; pp 509–522.

Merlo, L.; Ghisi, R.; Rascio, N.; Passera, C. Effect of Humic Substances on Carbohydrate Metabolism of Maize Leaves. *Can. J. Plant Sci.* **1991**, *71*, 419–425.

Mohammadi, K.; Ghalavand, A.; Aghaalikhani, M.; Heidari, G.; Shahmoradi, B.; Sohrabi, Y. Effect of Different Methods of Crop Rotation and Fertilization on Canola Traits and Soil Microbial Activity. *Aust. J. Crop Sci.* **2011**, *5*, 1261–1268.

Morales-Corts, M. R.; Gómez-Sánchez, M. A.; Pérez-Sánchez , R. Evaluation of Green/Pruning Wastes Compost and Vermicompost, Slumgum Compost and Their Mixes as Growing Media for Horticultural Production. *Sci. Hortic.* **2014**, *172*, 155–160.

Naguib, A. E. M. M.; El-Baz, F. K.; Salama, Z. A.; Hanaa, H. A. E. B.; Ali, H. F.; Gaafar, A. A. Enhancement of Phenolics, Flavonoids and Glucosinolates of Broccoli (*Brassica olaracea*, var. *Italica*) as Antioxidants in Response to Organic and Bio-Organic Fertilizers. *J. Saudi Soc. Agric. Sci.* **2012**, *11*, 135–142.

Nicolas, C.; Masciandaro, G.; Hernandez, T.; Garcia, C. Chemical-Structural Changes of Organic Matter in a Semi-Arid Soil after Organic Amendment. *Pedosphere* **2012**, *22*, 283–293.

Olivares, F. L.; Aguiar, N. O.; Rosa, R. C. C.; Canellas, L. P. Substrate Biofortification in Combination with Foliar Sprays of Plant Growth Promoting Bacteria and Humic Substances Boosts Production of Organic Tomatoes. *Sci. Hortic.* **2015**, *183*, 100–108.

Oyedeji, S.; Animasaun, D. A.; Bello, A. A.; Agboola, O. O. Effect of NPK and Poultry Manure on Growth, Yield, and Proximate Composition of Three Amaranths. *J. Bot.* **2014**. http://dx.doi.org/10.1155/2014/828750.

Pant, A. P.; Radovich, T. J. K.; Hue, N. V.; Paull, R. E. Biochemical Properties of Compost Tea Associated with Compost Quality and Effects on Pak Choi Growth. *Sci. Hortic.* **2012**, *148*, 138–146.

Pokrovsky, O. S.; Schott, J. Iron Colloids/Organic Matter Associated Transport of Major and Trace Elements in Small Boreal Rivers and their Estuaries (NW Russia). *Chem. Geol.* **2002**, *190*, 141–179.

Ponge, J. F. Plant-Soil Feedbacks Mediated by Humus Forms: A Review. *Soil Biol. Biochem.* **2013**, *57*, 1048–1060.

Prusty, B. A. K.; Azeez, P. A. Humus: The Natural Organic Matter in the Soil System. *J. Agril. Res. Dev.* **2015**, *1*, 1–12.

Pulido-Fernandez, M.; Schnabel, S.; Lavado-Contador, J. F.; Mellado, I. M.; Perez, R. O. Soil Organic Matter of Iberian Open Woodland Rangelands as Influenced by Vegetation Cover and Land Management. *Catena* **2013**, *109*, 13–24.

Ramana, S.; Biswas, A. K.; Singh, A. B.; Yadava, R. B. R. Relative Efficiency of Different Distillery Effluents on Growth, Nitrogen Fixation and Yield of Ground Nut. *Bioresour. Technol.* **2002**, *81*, 117–121.

Rastogi, M.; Singh, S.; Pathak, H. Emission of Carbon Dioxide from Soil. *Curr. Sci.* **2002**, *82*, 510–517.

Sahariah, B.; Sinha, I.; Sharma, P.; Goswami, L.; Bhattacharyya, P.; Gogoi, N.; Bhattachrya, S. S. Efficacy of Bioconversion of Paper Mill Bamboo Sludge and Lime Waste by Composting and Vermiconversion Technologies. *Chemosphere* **2014**, *109*, 77–83.

Schimel, J. P.; Schaeffer, S. M. Microbial Control over Carbon Cycling in Soil. *Front. Microbiol.* **2012**. http://dx.doi.org/10.3389/fmicb.2012.00348.

Schmidt, H.-P.; Kammann, C.; Niggli, C.; Evangelou, M. W. H.; Mackie, K. A.; Abiven, S. Biochar and Biochar-Compost as Soil Amendments to a Vineyard Soil: Influences on Plant Growth, Nutrient uptake, Plant Health and Grape Quality. *Agric. Ecosyst. Environ.* **2014**, *191*, 117–123.

Simpraga, M.; Verbeeck, H.; Demarcke, M.; Joo, E.; Pokorska, O.; Amelynck, C.; Schoon, N.; Dewulf, J.; Van Langenhove, H.; Heinesch, B.; Aubinet, M.; Laffineur, Q.; Muller, J.-F.; Steppe, K. Clear Link between Drought Stress, Photosynthesis and Biogenic Volatile Organic Compounds in *Fagus sylvatica* L. *Atmos. Environ.* **2011**, *30*, 5254–5259.

Singh, D.; Suthar, S. Vermicomposting of Herbal Pharmaceutical Industry Waste: Earthworm Growth, Plant-Available Nutrient and Microbial Quality of End Materials. *Bioresour. Technol.* **2012**, *112*, 179–185.

Singh, R.; Singh, R.; Soni, S. K.; Singh, S. P.; Chauhan, U. K.; Kalra, A. Vermicompost from Biodegraded Distillation Waste Improves Soil Properties and Essential Oil Yield of *Pogostemon cablin* (patchouli) Benth. *Appl. Soil Ecol.* **2013**, *70*, 48–56.

Taie, H. A. A.; El-Mergawi, R.; Radwan, S. Isoflavonoids, Flavonoids, Phenolic Acids Profiles and Antioxidant Activity of Soybean Seeds as Affected by Organic and Bioorganic Fertilization. *Am.-Eur. J. Agric. Environ. Sci.* **2008**, *4*, 207–213.

Tanase, C.; Boz, I.; Stingu, A.; Volf, I.; Popa, V. I. Physiological and Biochemical Responses Induced by Spruce Bark Aqueous Extract and Deuterium Depleted Water with Synergistic Action in Sunflower (*Helianthus annuus* L.) Plants. *Ind. Crop Prod.* **2014,** *60,* 160–167.

Toussoun, T. A.; Weinhold, A. R.; Linderman, R. G.; Patrick, Z. A. Nature of Phytotoxic Substances Produced during Plant Residue Decomposition in Soil. *Phyopathol.* **1967,** *58,* 41–45.

Trevisan, S.; Francioso, O.; Quaggiotti, S.; Nardi, S. Humic Substances Biological Activity at the Plant–Soil Interface: From Environmental Aspects to Molecular Factors. *Plant Sig. Behav.* **2010,** *5*(6), 635–643.

Tyler, G.; Olsson, T. Plant Uptake of Major and Minor Mineral Elements as Influenced by Soil Acidity and Liming. *Plant Soil.* **2001,** *230,* 307–321.

Varadachari, C.; Mondal, A. H.; Nayak, D. C.; Ghosh, K. Clay–Humus Complexation: Effect of pH and Nature of Bonding. *Soil Biol. Biochem.* **1994,** *26,* 1145–1149.

Virto, I.; Imaz, M. J.; Enrique, A.; Hoogmoed, W.; Bescansa, P. Burning Crop Residues under No-Till in Semi-Arid Land, Northern Spain—Effects on Soil Organic Matter, Aggregation, and Earthworm Populations. *Soil Res.* **2007,** *45,* 414–421.

Vivas, A.; Moreno, B.; Garcia-Rodriguez, S.; Benitez, E. Assessing the Impact of Composting and Vermicomposting on Bacterial Community Size and Structure, and Microbial Functional Diversity of an Olive-Mill Waste. *Bioresour. Tech.* **2009,** *100,* 1319–1326.

Wei, H.; Guenet, B.; Vicca, S.; Nunan, N.; Asard, H.; AbdElgawad, H.; Shen, W.; Janssens, I. A. High Clay Content Accelerates the Decomposition of Fresh Organic Matter in Artificial Soils. *Soil Biol. Biochem.* **2014,** *77,* 100–108.

Wheeler, B. E. J. The Conversion of Amino Acids in Soil. *Plant Soil.* **1963,** *19,* 219–232.

Xu, Z.; Yu, G.; Zhang, X.; Ge, J.; He, N.; Wang, Q.; Wang, D. The Variations in Soil Microbial Communities, Enzyme Activities and their Relationships with Soil Organic Matter Decomposition along the Northern Slope of Changbai Mountain. *Appl. Soil Ecol.* **2015,** *86,* 19–29.

Zandonadi, D. B.; Santos, M. P.; Busato, J. G.; Peres, L. E. P.; Facanha, A. R. Plant Physiology as Affected by Humified Organic Matter. *Theor. Exp. Plant Physiol.* **2013,** *25,* 12–25.

Zhang, L.; Sun, X.-y.; Yun, T.; Gong, X.-q. Biochar and Humic Acid Amendments Improve the Quality of Composted Green Waste as a Growth Medium for the Ornamental Plant *Calathea insignis. Sci. Hortic.* **2014,** *176,* 70–78.

CHAPTER 3

PLANT AND SOIL INTERFACES AND THEIR INTERACTIONS UNDER DIFFERENT CLIMATE

SUBHASISH DAS and SATYA SUNDAR BHATTACHARYA*

Department of Environmental Science, Tezpur University, Tezpur 784028, Assam, India

*Corresponding author, Tel.: +91 3712 267007/+91 3712 2670078/+91 3712 2670079x5610; E-mail: satyasundarb@yahoo. co.in; satya72@tezu.ernet.in.

CONTENTS

ABSTRACT

Plant life on earth depends inevitably on soil. They respond to the soil system in various ways which modifies their existence and very growth significantly. Soil comprises a complex physicochemical environment and harbors diverse biota that continuously interact with the atmosphere and vice versa. The soil fertility and productivity are very much dependent on the environment for its maintenance.

3.1   THE SOIL ENVIRONMENT

The soil environment in true sense can be defined as the synthesis of physical, chemical, and biological attributes (Passioura, 2002). A suitable environment is required for achieving better plant growth, productivity, and yield. Research throughout the world has shown that the soil system sends certain stimulus to the plants received through their roots and further relayed to the shoot which controls their ability of water and nutrient uptake (Davies & Zhang, 1991; Aiken & Smucker, 1996; Jackson, 1993). The favorable soil physical environment for plants is deep soil with high water-holding capacity, proper aeration, and low resistance. Availability of plant essential nutrients forms the desired soil chemical environment. The biological environment comprises beneficial soil organisms. The soil environment, unlike its aerial counterpart, can be managed sustainably for prolonged support (Reddy & Reddy, 2010).

3.1.1   SOIL FERTILITY AND PRODUCTIVITY

Soil fertility is the capacity of soil to provide nutrients to plants in sufficient amount. The soil productivity is the ability of soil to produce crops under certain management scheme. Soil fertility is intrinsically related to soil productivity (Sanchez & Leaky, 2014). All productive soils can be designated as fertile but all fertile soils may or may not be productive. Problems like water-logging, salinity, harsh climate, and others, hamper the soil productivity.

Soil fertility

- It can be regarded as measure of available plant nutrients.
- It forms one of the major factors for crop production.
- The fertility of soil can be calculated under laboratory conditions.
- It defines the status of soil toward crop production.

Soil productivity

- It implies to the crop yield.
- The productivity or yield is determined by the interaction of all soil environmental factors.
- This can only be ascertained in field conditions under particular climatic situations.
- It implies to the product of various factors influencing soil management.

3.1.2   FERTILITY LOSES AND MAINTENANCE

Degradation and depletion of soil fertility occur due to excessive removal of nutrients by crops, leaching, erosion, gaseous loses, and so on. Restoration of soil fertility is a big problem for modern agriculture. Maintenance of soil fertility should be considered both on a temporary and long-term basis. The temporary measures include suitable cultural practices such as addition of organic manures, green manures, biofertilizers, and fertilizers. The long-term measures are reclamation of the soils by amendments and adopting suitable soil conservation practices. Reports show that intercropping is a lucrative option to restore the fertility loss of soil. For example, cassava-legume intercropping has been shown to be highly beneficial for gaining high productivity in the Congo highlands (Pypers et al., 2011). Similarly, Stark et al. (2015) reported that incorporation of fast-growing tree species like *Betula* sp. and *Populus* sp. enhance the fertility and productivity of soil.

3.1.3   SOIL PHYSICAL ENVIRONMENT

Soil is a complex system which is comprised of three major heads: soil solids, soil solution (liquid), and gaseous phase that are arranged in a jumbled framework with significant solid–liquid, liquid–gas, and gas–solid interfaces (Wang et al., 2015). Soil physical environment is controlled by soil characters like texture, structure, aeration, water, mechanical resistance, and depth of soil. The union of the three phases poses a myriad influence on the heat storage ability, internal fluid mobility, root growth, gaseous exchange, extent of infiltration, and so on, in soil (Bachmair et al., 2009; Tokumoto et al., 2010).

3.1.3.1 SOIL TEXTURE

3.1.3.1.1 Primary Particles

Soil texture refers to the relative proportions of sand, silt, and clay. It is an important factor which influences the mineralization rate of carbon (C) and nitrogen (N) under a given set of temperature and moisture conditions (Craine & Gelderman, 2011; Harrison-Kirk et al., 2014). Generally, rock fragments larger than 2 cm in diameter are stones, materials between 2-cm and 2-mm diameter are gravels, and the soil mineral matter smaller than 2-mm diameter is the fine earth. Sand, silt, and clay together constitute the fine earth. The primary soil particles are elaborated in Table 3.1.

TABLE 3.1   Soil Primary Particles.

Particle	Diameter (mm)	Reference
Clay	<0.002	Brady and Weil (1999)
Silt	0.02–0.002	
Fine sand	0.2–0.02	
Coarse sand	0.2–2	
Fine earth	<2	
Gravel	2.20	
Stone	>20	

Source: Reddy, Y. L.; Reddy, G. H., 2010. *Principles of Agronomy*. Kalyani Publishers: New Delhi.

3.1.3.1.1.1 Sand

The sand particles of 0.2–2 mm in diameter are classified as coarse sand and of 0.02–0.2 mm in diameter as fine sand. Sand particles are small pieces of unweathered rock fragments. Unless these particles are coated with clay or silt, they donot exhibit properties such as plasticity, cohesion, stickiness, moisture, nutrient retention, and so on. Because of the large size of sand particles macropores exist between them which facilitate free movement of air and water.

3.1.3.1.1.2   Silt

The size of the silt particles are in the range 0.02–0.002 mm in diameter. Because of an adhering film of clay, they exhibit some plasticity, cohesion, adhesion, and adsorption. These silt particles can hold more amount of water than sand but less than clay. Both silt and sand particles are approximately spherical and cubical in shape.

3.1.3.1.1.3   Clay

Clay fraction is the most important constituent of soil that controls most of the soil physical and chemical properties. These particles are of size <0.002 mm in diameter. They have the highest surface area since surface area is inversely related to size. They can adsorb and retain water and nutrients. Some clays swell on wetting and shrink on drying. They exhibit properties like flocculation (clustering), deflocculation, and plasticity. Clay behaves like a weak acid which is neutralized by bases such as Ca^{2+} and Mg^{2+} ions, thus serving as store house for several nutrients.

3.2   TEXTURAL CLASSES

Soil texture is classified based on the proportion of predominant size fraction of sand, silt, and clay. It is a prominent soil attribute that governs the nutrient transformation in soil. Rajaie et al. (2008) reported the Ni-transformation with respect to the textural classes in calcareous soil of Iran. If the soil contains more than 80% of silt fraction, the soil fraction is designated as silty soil. Soil with more than 85% of sand is called sandy soil and soil with 40% clay as clay soil. If all the three fractions are in sizeable proportion, the soil is called as loamy soil. Silty clay means a soil in which the clay characteristics are outstanding and which also contains sufficient silt. Ranges of sand, silt, and clay percentages for each textural class are presented in Table 3.2. Textural classes of a soil can be known from the textural triangle (Figure 3.1).

TABLE 3.2   Soil Textural Classes.

Textural class	Ranges (%)			Reference
	Sand	**Silt**	**Clay**	
Clay	0–40	0–40	40–60	Brady and Weil (1999)
Silty clay	0–20	40–60	40–60	
Sand	40–65	0–20	35–45	
Sandy clay	85–100	0–15	0–10	
Loamy sand	70–90	0–30	0–15	
Sandy loam	43–80	0–50	0–20	
Loam	23–52	28–50	7–27	
Silt loam	0–50	50–88	0–27	
Silt	0–20	88–100	0–12	
Sandy clay loam	45–80	0–28	20–55	
Clay loam	20–45	15–53	27–40	
Silty clay loam	0–20	40–73	27–40	

Source: Reddy, Y. L.; Reddy, G. H., 2010. *Principles of Agronomy*. Kalyani Publishers: New Delhi.

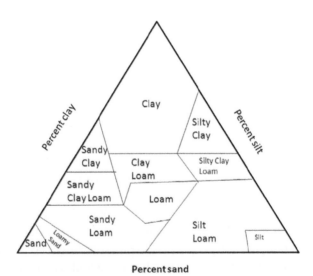

Percent sand

FIGURE 3.1   Textural triangle for determining soil texture.

3.2.1 SIGNIFICANCE OF SOIL TEXTURE

Soil texture is a prominent feature of soil system that affects the physical, chemical, and biological properties of soil, thereby modifying soil to a greater extent (Lamsal & Mishra, 2010). Soil texture is a permanent feature of soil and its change over years is negligible. It can be changed by adding sand or silt or clay as an amendment to improve physical condition. For example, tank silt is added to sandy soil to improve the water-holding and nutrient retention capacities (Vaidyanathan, 1992). Sand is added to heavy clay soil to improve internal drainage. But these operations are expensive. Application of rice husks at 4.5 t/ha decreased soil bulk density, increased saturated hydraulic conductivity. Seed yield and plant height of maize increased significantly with improvement of soil physical properties. White grub damage is less in crops grown in clay soil compared to loamy and sandy soils.

Long-term application of small doses of NPK (40–100 kg) contributes to a biological transformation of arable horizons and forms a high quality of biogenic pores and aggregates. Improvement of water permeability and drainage caused removal of the organomineral compounds from the upper horizons. Optimum doses of NPK applied to form biogenic aggregates were N 120 P 120 K 120.

Soil texture influences soil physical and chemical properties like water-holding capacity, nutrient retention, nutrient fixation, nutrient availability, drainage, strength, compressibility, and thermal regime. The suitability of a soil to a particular crop depends on texture in addition to soil depth, depth of water table, salinity, and alkalinity.

3.3 SOIL STRUCTURE

The soil particles are grouped into definite structures called the soil aggregates. Soil structure is the arrangement of the different soil particles into a certain defined pattern involving multishaped and -sized pores formed by the action of several actions over a period of time (Letey, 1991). Aggregate formation and strength depends on swelling and shrinkage processes and on biological activity and kinds of organic exudates as well as on the intensity, number, and time of swelling and drying events. The intra-aggregate pore distribution is vital for a stable soil structure which is important for aeration and gaseous composition of the intra-aggregate pore space. The gaseous composition in the soil aggregate pores depends highly on the amount of dissolved organic matter or exudates. Important research on soil structure answers the questions related to the ability of soil to store C, solute mobility across the soil phases, and various organomineral complexing phenomena (Peng et al., 2015).

From the mechanical point of view, the strength of single aggregate, determined as the angle of internal friction and cohesion, depends on the number of contact points or the forces, which can be transmitted at each single contact point. The more structured soils are, the higher the proportion of the effective stress on the total stress is, but even in single aggregates positive pore water pressure values can be revealed. The high-tensile property of soil aggregation imparts it a solid interface for efficient coupling with mechanical structures built over it (Aldaikh et al., 2015).

3.3.1 SOIL CHEMICAL ENVIRONMENT

Soil chemical environment is a complex system which is aided by different reactions and interactions between the soil attributes. The dynamics of nutrient transformation and plant nutrient availability largely depends on pH, clay minerals, and cation (CEC) and anion exchange capacity (AEC). The soil chemical environment is most vulnerable to erratic changes when shifting from traditional to more fertilizer-driven mode of agriculture. As such, these alterations in agrotechniques trigger a series of odd processes which hinder the normal soil chemistry leading to poor crop yield (Clark et al., 1998).

3.3.1.1 SOIL pH

The pH is also referred to as the soil reaction and is defined as the negative logarithm of hydrogen ion activity. The pH of a soil indicates its acidity or alkalinity. pH influences nutrient availability, soil physical conditions, and plant growth (Jiang et al., 2015). The rate of nutrient release through its influence on decomposition, CEC, solubility of materials, and others are also mediated through a change in soil pH. Decomposition of organic matter is slowest at pH 6 and fastest at pH 6–8. Availability of Fe, Al, Mn, Ca, and Zn are influenced by pH through its effect on solubility of their nutrient compounds. The optimum pH range for availability of different nutrients can be found in Table 3.3.

TABLE 3.3   Optimum pH for Nutrient Availability.

Nutrient	Optimum pH range	Reference
N	6.0–8.0	Brady and Weil (1999)
P	6.5–7.5	
K	6.0–7.5	
S	6.0 and above	
Fe	6.0 and above	
Mn	5.0–6.5	
Bo, Cu, Zn	5.0–7.0	
Mo	7.0 and above	
Ca and Mg	7.0–8.5	

Source: Reddy, Y. L.; Reddy, G. H., 2010. *Principles of Agronomy*. Kalyani Publishers: New Delhi.

Moreover, soil physical conditions are related to soil pH. A soil with pH above 8.5 abounds in Na^{2+} content that hastens the deflocculation of soil colloids resulting in destruction of soil structure. Thus, pH has an indirect role in plant growth regulation through its effect on nutrient availability and soil physical conditions. Table 3.4 shows the preferred pH for different crops.

TABLE 3.4   Soil pH Preferences for Different Crops.

Crop	pH	Crop	pH	Reference
Cotton	5.0–6.5	Soyabean	5.5–7.0	Reddy and Reddy (2010)
Potato	5.0–5.5	Rice	4.0–6.0	
Tea	4.0–6.0	Wheat	6.0–7.5	
Oats	5.0–7.5	Barley	5.0–6.5	
Sugarcane	6.0–7.5			
Sorghum	6.0–7.5			
Maize	6.0–7.5			
Pea	6.0–7.5			
Millets	5.5–7.0			

Source: Reddy, Y. L.; Reddy, G. H., 2010. *Principles of Agronomy*. Kalyani Publishers: New Delhi.

3.3.1.2   CLAY MINERALS

A mineral can be defined as a naturally occurring inorganic material with a definite chemical composition and specific physical characters. A mineral that forms the original component of a rock is known as primary mineral, for example, feldspar, mica, and others. A mineral that has been formed as a result of certain changes in rocks throughout a period of time is known as secondary mineral, for example, kaolinite, montmorillonite, illite, and others. A magnified image of the kaolinite and montmorillonite clay particles is presented in Figure 3.2. The hydrated aluminosilicate secondary minerals with particle size <0.002 mm in diameter are called clay minerals. Clay minerals act as active reservoirs of various cations which are of great importance to plant life (Barre et al., 2008). The clay minerals together with organic matter constitute the colloidal fractions of soil. Colloids are particles of size <0.0002 mm in diameter. The individual particle of a colloid is termed as a micelle. There are two types of soil colloids, namely, inorganic and organic colloids. The inorganic colloids may either be silicate or non-silicate clay minerals.

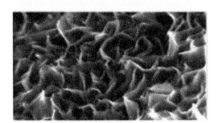   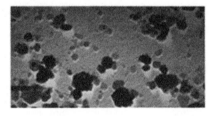

Montmorillonite   **Kaolinite**

FIGURE 3.2   Clay mineral types: montmorillonite and kaolinite under high magnification.

3.4   SILICATE CLAY MINERALS

The three important silicate clay minerals are kaolinite, montmorilonite, and illite. Depending on the number of silica and alumina layers, they are classified as 1:1 and 2:1 clay minerals (Martin et al., 1991).

3.4.1   KAOLINITE GROUP

Each crystal unit of kaolinite group consists of one sheet each of silica and alumina. Hence, these are 1:1 type of clay minerals. These two sheets are held together by oxygen atoms which are mutually shared by Si and Al atoms of their respective sheets. Due to the tight bonding pattern, expansion is not possible which makes them impermeable to water and other cations. Kaolinite minerals are bigger in size ranging between 0.1 and 5.0 μm in width.

3.4.2   MONTMORILLONITE GROUP

The montmorillonite micelle belongs to 2:1 type with two sheets of silica and one sheet of alumina. These sheets are held loosely by weak van der Waal's forces making them expandable allowing water molecules and cations inside the geometry. The size of these crystals ranges between 0.1 and 1.0 μm.

3.4.3   ILLITE GROUP

The illite group mainly consists of 2:1 type architecture. These micelles are held by K^+ ions more tightly than oxygen linkages as in montmorillonite. These micelles are less expandable than montmorillonite. Figure 3.3 portrays the expandable and non-expandable clay particles with ionic distribution.

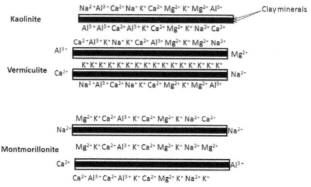

FIGURE 3.3   Schematic diagram of charged clay minerals.

3.5   NON-SILICATE CLAY MINERALS

3.5.1   HYDROUS OXIDES OF IRON AND ALUMINUM

The red and yellow soils of tropical and subtropical regions abound in the hydrous oxides of iron and aluminum. These oxides have different number of water molecules, depending on the type of clay mineral. For example, gaethite is an iron oxide with one water molecule and gibbsite is a hydrous aluminum oxide with three water molecules in its structure. Most hydrous oxides are less cohesive than silicate clays. They also carry less negative charges than silicate clays. A much better physical condition is imparted to the soil by these metal-hydrous oxides.

3.5.2   ALLOPHANES

Amorphous mineral matter in soil is called allophones which are combination of silica and sexquioxides. They have an approximate composition which is found in many soils. Allophanes have high cation and anion exchange capacities.

3.6   ORGANIC SOIL COLLOIDS (HUMUS)

Humus is the end-product of decay and decomposition of organic residues, amorphous, dark brown in color and is quite stable in nature. Organic colloids show remarkable water retention capacity and a better cation adsorption capacity when compared to inorganic colloids. Major research in the chemistry and structure of soil organic colloids reveals its importance in pollutant transport and heavy metal chelation (Zirkler et al., 2012; Wang et al., 2014).

3.6.1   CEC

CEC is a very important property of the soil when plant nutrition is considered (Abdul Khalil et al., 2015). Most clay micelles have a crystalline structure while humus micelles are amorphous. These micelles carry negative charges on their surface due to isomorphic substitution, ionization of hydroxyl groups, and exposed carbonyl and hydroxyl groups. Isomorphic substitution of ions results mostly in higher negative charge. The CEC of soil greatly inter-related to the availability of soil organic matter (Bradl, 2004). Due to the presence of negative charge, cations are adsorbed on the surface of the micelles. The micellar structure holds the cations in two layers, making it a double-charged layer. An illustration of CEC is provided in Figure 3.4.

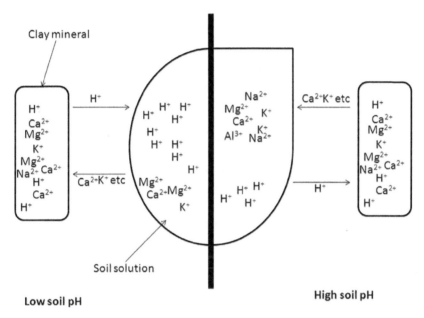

FIGURE 3.4   Schematic diagram of cation exchange capacity in soil.

CEC of soils influences the capacity of the soil to hold nutrients such as Ca, Mg, NH_4^+, and so on, and the quantity of a nutrient required to change its relative level in soil. Percent base saturation indicates the proportion of basic cations in CEC. The plant nutrient availability highly depends on the percent saturation of the cations in soil.

3.6.2   AEC

AEC is less in magnitude compared to CEC. The sites for anion exchange in soil colloids are present at the amine groups in humus and at the edge of a silicate clay mineral when the bonding between clay mineral and cation terminates. Release of OH⁻ from Fe and Al causes positive charges on clay minerals. Moreover, several organic anions are also produced by the microbial decomposition of organic matter and plant exudates that have significantly high affinity for cations in the surrounding media (Nguyen et al., 2013). Thus, due to the difference in the charges, the anions are exchanged between the soil colloids and the soil solution.

3.6.3 SOIL SOLUTION

Soil water content is an important aspect which governs the type and availability of vegetation in a given place (Misra & Tyler, 1999). Soil consists of a pool of dissolved nutrients called the soil solution. The solution is in dynamic equilibrium with the clay particles. The solubility of a material determines its ability to move in soils. Soluble salts are chlorides, sulfates, nitrates, and some carbonates and bicarbonates of Na, K, Mg, Ca, and NH_4 (Larsen & Widdowson, 1968). When a material is dissolved in water, they dissociate into cations and anions (Figure 3.5). Water molecules rearrange around these ions as they are dipole and form shells around the ions. These ions, especially cations, are easily available to plants. The concentration of soil solution and its pH are changing continuously. Soil solution and clay minerals exchange cations depending on addition or depletion from soil solution. When fertilizers are added, soil solution acquires more cations and more of these ions are adsorbed in the clay surface. When nutrients are adsorbed by plants, cations from clay complex move to the soil solution. The pH of soil solution varies with the crop growth stage. When plants adsorb cations, H^+ is released into the solution making it acidic and the opposite occurs when anions are adsorbed. Rainwater or irrigation dilutes the soil solutions while dry periods increase the concentration. The minimal shift in the soil pH does not affect the soil due to its inherent capacity called the soil buffering capacity. Modern research has shed light on some other important aspects of soil solution. McDowell and Liptzin (2014) reported soil solution chemistry to be indicative of hurricanes with significantly different concentration of ions in the soil solution to pre-and post-hurricane events.

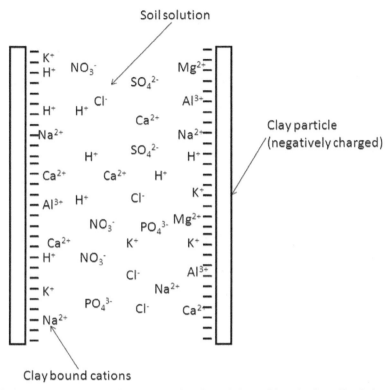

FIGURE 3.5 Schematic diagram showing flux of charged ions in the soil solution and clay interfaces.

3.7 SOIL–PLANT–CLIMATE RELATIONSHIP

The soil–plant–climate relationship is a continuous cyclic process that operates between the soil, its vegetation, and the surrounding environment. This is a very complex but important relation that plays an integral part in growth and development of the plant, replenishment of the soil losses, and modification of the climate (Krumins et al., 2015). Water movement in the atmosphere follows the same principles. The free-energy level of water is the key factor that determines the soil–water behavior. The same can be said for soil–plant and plant–atmosphere relations (Figure 3.6). As water moves through soil to plant roots, into the roots, across cells into stems, up the plant stems to the leaves, and is evaporated from the leaf surfaces, its tendency to move is determined by differences in free energy of the water or by the moisture potential (Sheil, 2014). The moisture potential must be higher in the soil than in the plant roots if water needs to be taken up from the soil. Likewise, movement up the stem to the leaf cells is in response to differences in moisture potential. At the community level, the physiological profiles of plants and microbes are influenced highly by the climate–plant–soil interactions (Pailler et al., 2014).

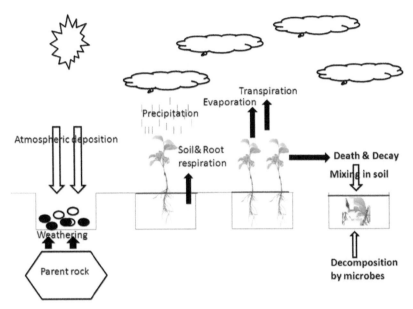

FIGURE 3.6   Soil–plant–climate continuum (schematic).

3.7.1   EVAPOTRANSPIRATION

The vapor losses of water from soils occur by evaporation at the soil surface and by transpiration from the leaf surfaces. The combined loss resulting from evaporation and transpiration is termed as evapotranspiration. The evapotranspiration is responsible for most of the water removal from soils during a crop-growing period (Jasechko et al., 2013). The phenomenon is of special significance to growing plants. Some plant characteristics can affect evapotranspiration over a growing season. The length of the crop growth period will have an influence. Likewise, the depth of rooting will determine the subsoil moisture available for absorption and eventually for transpiration. These interactions illustrate the importance of such relationships with respect to water conservation. A typical evapotranspiration condition is given in Figure 3.7.

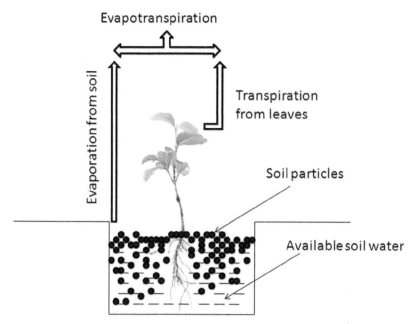

FIGURE 3.7   Schematic diagram showing evapotranspiration.

There is a marked variation in the evapotranspiration of different climatic areas (Vicente-Serrano et al., 2014; Zhang et al., 2015). A number of factors determine the relative water losses from the soil surface and from transpiration:

1. Plant cover in relation to soil surface (leaf area index);
2. Efficiency of water use by different plants;
3. Proportion of time the crop is on land, especially during the summer months; and
4. Climatic conditions.

Loss by evaporation from soil is generally proportionately higher in drier regions than in humid areas. Such vapor loss is at least 60% of the total rainfall for dry land areas. Losses by transpiration account for about 35%, leaving about 5% for run-off.

3.7.2   EFFICIENCY OF WATER USE

Water use efficiency can be defined as the amount of yield obtained per unit of applied water (Howell, 2003). The crop production from the use of a given amount of water is an important aspect especially in areas where moisture is scarce (Rahil & Qanadillo, 2015). The ecosystem water use efficiency links both the ecological and hydrological processes and sources coordination between the carbon sequestration and water usage in a particular soil set up (Li et al., 2015). Water availability in soil is important because certain imbalance in root water intake and transpirational moisture loss gives rise to dehydrated or shrunken plant tissue (Aroca et al., 2012). On the contrary, excess of soil moisture may obstruct the movement of oxygen and carbon dioxide in the plant tissue resulting into poor photosynthesis (Muilu-Makela et al., 2015). The efficiency can be discussed under the following two heads.

3.7.2.1   TRANSPIRATION EFFICIENCY

The transpiration efficiency can be expressed as dry matter yield per unit of water transpired. The transpiration efficiency for any given crop is markedly affected by climatic conditions and soil factors (Wu et al., 2015). There are differences in transpiration efficiency of different plant species under the same climatic conditions. The cereal crops have high transpiration efficiency compared to legumes while crops like potato, oats, and barley have intermediate transpiration efficiency.

3.7.2.2   EVAPOTRANSPIRATION EFFICIENCY

Evapotranspiration efficiency is a more variable process because it depends both on evaporation and transpiration from the soil surface. It is an important ecophysiological attribute which is a latent heat flux and maintains the water and energy balance (Xiao et al., 2013). Furthermore, certain soil and crop management practices that influence evaporation from soil can affect evapotranspiration and, in turn, evapotranspiration efficiency. The evapotranspiration efficiency is affected by climate, being higher in moist areas than in dry areas. Evapotranspiration efficiency also responds to factors that drastically affect crop yields. Highest efficiency is attained where moisture and nutrient status are generally satisfactory for crop production.

3.7.3   CONTROL OF WATER INFILTRATION

Water infiltration is a natural phenomenon which is mainly governed by the soil's physical characteristics (Elmashad & Ata, 2014). The efficient use of water is affected by two types of management practices:

1. Practices that increase the amount of water entering the soil, and
2. Practices that increase the crop production per unit of water taken up by plants.

The management practices that keep the soil surface receptive to water penetration enhance the proportion of rain or irrigation water entering the soil. This also covers the soil surface and protects it from the beating action of rain drops. Included are selected tillage and engineering practices, especially those that leave considerable residues on the surface. In some areas, water infiltration and root penetration are constrained by subsurface layers and pans. Special efforts are needed to break up these layers. Simultaneous attempts have been made to introduce chopped organic residues into the subsoil layers (vertical mulching) and placing the fertilizers into the deep layers of soils. While increased water infiltration and some yield increases have been achieved using these practices, their high costs have restricted their general use.

3.7.4   CONTROL OF EVAPOTRANSPIRATION

Soil and crop management has a crucial role in the control of evaporation from soil surface (Zha et al., 2013; Brown et al., 2014). Evapotranspiration control is strictly based on the principles that tend to reduce transpiration. Thus, major focus should be on those practices that reduce evaporation, so as to maximize the water remaining for transpiration. Soil management considerations will follow those concerning crop management.

3.7.4.1   CROP SELECTION AND MANAGEMENT

Both transpiration and evapotranspiration rely on the crop species and crop management practices. The selection of a crop species is a crucial step that checks both transpiration and evapotranspiration (Figure 3.8). For

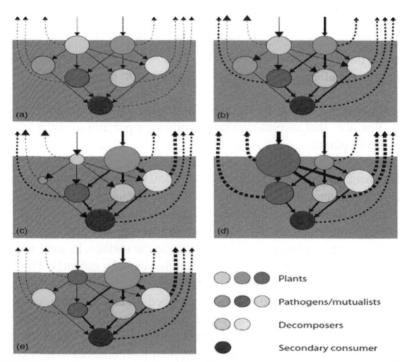

FIGURE 3.8 Changes in trophic relationships as per hierarchy of responses to climate change and carbon cycling. Arrows indicate flow of carbon (solid arrows: net input; dashed arrows: net output, and size of arrows are proportional to flow). Circle sizes indicate relative abundance of species in a simplified soil food web, and colors indicate species identity. (a) The undisturbed system prior to the initiation of chronic global change forces. (b) The system after the initiation of individual effects but before community reordering with one species less able to perform under altered conditions. (c) The system after community reordering has occurred with the poorly performing plant species becoming less abundant in competition with responding species. The abundance of its specialist symbiont or pathogen is proportionately reduced, whereas the abundance of one decomposer increases with increasing production of the responding plant species. (d) A long-term response in which the poorly performing plant species, and its pathogen or symbiont, is lost from the system, and in which a new competitively superior plant species is added that has escaped its natural enemies. As a result of the introduction of this highly successful species, the biomass of the non-specialist mutualist or pathogen increases, and the biomass of one decomposer remain high. (e) A long-term response in which an invasive microbe reduces the abundance of the invasive plant species, thus increasing the competitive ability of the native plant species and dampening the contribution of decoupling of ecosystem carbon cycling. (*Source*: Bardgett et al., 2013. Hierarchical Responses of Plant–Soil Interactions to Climate Change: Consequences for the Global Carbon Cycle. *J. Ecol.*, *101*, 334–343; Used with permission from John Wiley & Sons.)

example, corn and sorghum have lower water requirements than alfalfa, and the requirement of wheat and other cereals is generally intermediate. Also, when possible, growing a crop during cool season decreases both transpiration and evapotranspiration as vapor pressure gradients are low in those times (Brady, 1995).

3.7.4.2   FALLOW CROPPING

Farming systems that alternate fallow 1 year with traditional cropping the next year are useful to conserve soil moisture. The main objective is to eliminate transpiration every 1 year, thereby increasing the soil moisture storage for the next crop. This is an effective system as it has been observed that moisture levels at planting time are commonly increased by fallow cropping and yields have generally been increased through its use (Manalil & Flower, 2014).

3.7.5   CONTROL OF SURFACE EVAPORATION

Most of the precipitation on earth's surface is usually returned to the atmosphere by evaporation. The evaporation losses can be controlled by management practices that are designed to check evaporation by providing some cover to the soil. This cover can best be provided by mulches and by selected conservation tillage practices (Verburg et al., 2012).

3.7.5.1   MULCHES

Mulch can be defined as any material applied at the surface of a soil mainly to reduce evaporation or to keep weeds down. Examples are saw dust, manure, straw, leaves, crop residues, and so on. Mulches are very useful in controlling evaporation (Shukla et al., 2014).

3.7.5.2   CROP RESIDUE AND CONSERVATION TILLAGE

Conservation tillage practices that adopt usage of the previous year's crop residue form one of the most promising soil management practice these days. These practices markedly reduce both vapor and liquid losses (Odhiambo & Irmak, 2012). Conservation tillage practices are of different kinds like stubble mulch tillage and no tillage. In stubble mulch tillage, the residues from previous crop are spread on the soil surface followed by tilling the land, whereas, in no tillage system, the new crop is planted directly in the residues of the previous crop with no plowing.

3.7.6 PERCOLATION LOSSES OF WATER

Percolation is a hydrological process that pertains to the downward movement of water from the soil surface. When a soil receives more water than its water retention capacity, water seeps down through the soil layers, and percolation losses occur (Kukal & Sidhu, 2004). Percolation loss implies to the movement of water into the deep layers from where the plant roots are unable to absorb water. Percolation losses are influenced by various factors like the amount of rainfall and its distribution, by run-off from the soil, by evaporation, by the character of the soil, and by the crop.

3.7.6.1 PERCOLATION–EVAPORATION BALANCE

The relationship among precipitation, run-off, soil storage, and percolation varies with the climate regime of a place. In the humid-temperate region, the rate of water infiltration into the soils is commonly greater, at some time of year, than the rate of evapotranspiration. Percolation occurs when the soil-field capacity is reached.

During winter and early spring months, percolation is highest due to low evaporation losses. During the summer, little percolation occurs because evapotranspiration exceeds the infiltration followed by depletion of soil water. Normal plant growth is possible only because of moisture stored in the soil the previous winter and early spring. The general trends in the temperate zone semi-arid region are the same as for the humid region. Soil moisture is stored during the winter months and the moisture deficit is fulfilled during the summer months. However, due to low rainfall, very meager amount of percolation occurs. Ultimately, transpiration losses occur when plant roots absorb the percolating water. It has been recently reported that soil management practices, like incorporation of rock fragments, selective intercropping, and others, have a powerful impact on the evaporation–percolation balance in soil system (Zhongjie et al., 2008; Wang et al., 2012).

3.7.7 LEACHING OF NUTRIENTS

Leaching of plant's essential nutrients is an important phenomenon that requires special attention. Climatic factors as well as soil–nutrient interactions are the key factors that influence the nutrient dynamics in soil (Long et al., 2015). In regions where water percolation is high, the potential for leaching is also high. In such areas, percolation of excess water occurs that hastens nutrient removal. In case of unirrigated semi-arid areas, very less percolation occurs which aids in lower nutrient loss through leaching.

3.7.7.1 NUTRIENT–SOIL INTERACTION

The dynamics of nutrient losses significantly vary with soil properties (Mendoza et al., 2014). Because of the higher rate of percolation and lower nutrient absorbing capacity, sandy soils are prone to nutrient loss than the clay soils. For example, soluble phosphorus is quickly bound chemically in fine-textured soils with appreciable amounts of Fe and Al oxides. Consequently, very little phosphorus is lost by leaching. Fe and Al hydrous oxides efficiently bind the sulfates and nitrates on their surfaces due to their high positive charges. For this reason, sulfates and nitrates are less prone to leach from soil rich in Fe oxides compared to soils where Fe oxides are less prominent.

CEC of soil positively influences the leaching of cations added in fertilizers. Soils with a high CEC tend to hold the added nutrients and prevent their leaching (Lehmann et al., 2003). Moreover, such soils are often naturally high in exchangeable cations, and thus a small portion of these exchangeable ions is continually subject to leaching loss.

3.7.7.2 SOIL AND WATER POLLUTION

Leaching losses pose several threats to the ecosystem functioning. Of these, there are two prime reasons for concern over the loss of nutrients from soil. The first concern is to find ways for keeping these nutrients in the soil so that essential elements are available to crop plants. The second concern involves certain management interventions that keep the nutrients from entering into the streams, rivers, and lakes (Grondona et al., 2014; Quan et al., 2014). Eutrophication or the excessive growth of algae and other hydrophytes takes place in water over enriched with N P K and other nutrients. This leads to depletion of dissolved oxygen in water posing serious threats to the aquatic life. Moreover, problems like ground water pollution and contamination of surface water due to run-off also arise if the nutrients leach more into the water systems.

3.8   INFLUENCE OF CLIMATE ON PLANTS

Climate over a place greatly influences the vegetation of the same. These interactions play a major role in determining the impact of climate change on ecosystem functioning and the carbon cycle involving mechanisms that operate over a wide range of spatial and temporal scales. Climate change mainly impacts the biogeochemical cycles which indirectly influence the above-ground and below-ground communities (Bardgett et al., 2013). These mechanisms operate across several spatial and temporal scales. The short-term effects appear over intra- and inter-annual timescales documenting climate change impacts on the physiology and activity of above-ground and below-ground biota. The long-term effects happen over tens to hundreds of years, when climate change can cause species range expansions and contractions. While climate change can impact directly on soil carbon, for instance by stimulating rates of decomposition under warming, there is also potential for indirect impacts via changes in vegetation. There is significant evidence that supports the fact which states that with climate-induced changes in the growth and structure of plant communities, the amount and quality of plant-derived carbon entering soil via root exudates and plant litter (shoot and root) is altered which again will have cascading effects on soil organisms and the processes of carbon cycling. Plants respond differently with the changing climate by upregulating or downregulating their physiology (Atkin & Tjoelker, 2003; Wu et al., 2011). The plant–climate interaction can be defined as a complex framework which deals with the results of climate-induced changes in plant–soil feedbacks for the carbon cycle. This framework involves a temporal and spatial hierarchy of plant–soil feedback responses to climate change and can be regarded as a tool for understanding the fate of these responses for the carbon cycle at local and global scales. It is based on the similar hierarchical response framework which addresses the issues of ecosystem responses to chronic resource alterations resulting from global change (Smith et al., 2009). This framework incorporates a hierarchy of mechanisms that will document the impact of climate change on ecosystem carbon dynamics at three levels of response, namely, individual and community reordering and species immigration and loss.

3.8.1 THE MECHANISM HIERARCHY

3.8.1.1 INDIVIDUAL PLANT RESPONSE TO CLIMATE

A diverse interaction takes place between plant individuals and climate. The latter modifies the physiology of the former giving rise to a series of ecophysiological responses (Korner et al., 2005). Some of the common responses involve changes in the activity, metabolism, behavior, and phenology of organisms, without alteration of community structure. At this scale, plant and soil communities are tightly coupled via mutual responses to environmental factors like temperature and moisture, which are again influenced by short-term variability in weather and climatic attributes. Such coupling is effectively instantaneous (Figure 3.8) and powered by plant inputs (litter and root exudates) of carbon and nutrients to soil. These inputs then act as a substrate for microorganisms which again regulate nutrient availability to plants through a series of mechanisms (Yuste et al., 2007). Mostly, changes in climate pattern affect the underlying plant–soil interactions by altering the magnitude of such networking. For example, increase in root exudation coupled with enhanced microbial activity significantly influences the carbon cycling. It is also known that plants respond to temperature rise and higher precipitation patterns by enhancing or downregulating the photosynthetic and respiration in order to maintain a positive carbon balance in the ecosystem. An increase in the atmospheric CO_2 concentration also stimulates plant growth and the availability of photosynthate within the plant, which can multiply the magnitude of carbon flux to roots in the form of root exudation of easily degradable sugars, organic acids, and amino acids (Jackson et al., 2009).

Such a degree of rhizodeposition in turn stimulates root respiration followed by soil organic matter mineralization that eventually leads to soil carbon losses. Moreover, the rate of rhizodeposition also triggers the rate of nitrogen mineralization that promises a sustained plant growth (Drake et al., 2011). It is also noteworthy that in spite of its positive side, rhizodeposition may also be deleterious in certain aspects. For example, in certain situations, an enhanced root carbon supply, coupled with an increased carbon-to-nitrogen ratio of plant litter, under elevated CO_2 can lead to nitrogen immobilization thereby limiting nitrogen availability to plants which creates a negative feedback on plant growth and carbon transfer to soil (Cotrufo et al., 1994).

Apart from the above-ground organisms, climate change also influences the activity of below-ground organisms. For example, atmospheric warming can directly increase soil microbial activity and promote the expression of certain genes that are involved in soil carbon degradation. However, it has been popularly debated that such an increase in microbial activity is temporary and in the real long-term situations, microbial respiration is very much tuned to increased temperatures. Moreover, warming can also affect microbial processes involved in nitrogen cycling which incorporates nitrogen mineralization, nitrification, and nitrogen fixation, resulting in increased plant nitrogen use and plush growth.

These indirect effects of warming are thought to be the primary factors that influence greater plant productivity under increased temperatures. However, there is little information available on the downregulation of plant or microbial processes under prolonged warming. Climate change can also decouple interactions between plant and soil communities. Decoupling can be defined as the disruption of a previously existing ecological interaction due to changes in activity or unavailability of either or both the plant and soil communities arising because of differences in phenology, abundance, dispersal, or extinction (Bardgett et al., 2002). In the short term, phenological differences will be the main cause of this decoupling. For example, changes in climate can affect plant phenology and alter both growing season and flowering patterns. There is some evidence that phenological changes above ground and below ground are closely tied together. For example, fungi which exhibit a diverse spatial preference are found both above- and below-ground conditions. Thus, fungi display species-specific phenology and time-lags in their phenological response, with fruiting responses being influenced by the previous year's weather conditions (Gange et al., 2007). These complex patterns point to the potential for phenological decoupling between above-ground and below-ground organisms, particularly if weather becomes more variable in the future.

Interruption in the network between plants and soil microbes related to nutrient supply can arise in a variety of short-term temporal scales that may hamper the carbon cycling in the long run. Such a situation is evident in alpine ecosystems where nitrogen is partitioned between plant and microbial communities over the growing season. Nutrient partitioning leads to N immobilization in winter when plant N demand falls and bacterial communities, which mineralize nutrients for plant use, thrive in the summer. A change in climate pattern possibly disrupts this intimate partitioning of nutrients between plant and microbial communities through increased soil freezing, as a result of reduced snow cover, thereby affecting the winter microbial communities and the dynamics of carbon and nitrogen cycling. In general sense, the below-ground organisms are comparatively more susceptible to extreme weather events. However, information on the relative response of below-ground and above-ground organisms to extreme weather events is absent which makes it difficult to assess whether or not they respond in synchrony to such environmental change. If mortality in response to extreme events of drought, heat, and freezing differs between below-ground and above-ground organisms, then decoupling will occur, with likely impacts on the carbon cycle.

3.8.2 CHANGES IN PLANT COMMUNITY STRUCTURE AND CLIMATE

The short-term changes in climate like changes in precipitation pattern, hike in temperature, and increase in atmospheric CO_2 concentration may fuel the reordering of both above- and below-ground individuals, thereby modifying the total community structure in the long run. The mechanism involves alteration in species abundance with no possible extinction or invasion of species. The alteration in plant community structure influences the amount and quality of organic carbon entering soil and modifies the soil physical and chemical environments (Bardgett et al., 2013). Therefore, climate-induced vegetation shifts can have substantial impacts on soil communities and carbon cycling. Research on soil–plant interaction shows that reduced precipitation and increased temperature aid for the settlement and growth of deeper rooting, woody plant species, which in turn increase below-ground carbon inputs and mycorrhizal colonization, thereby increasing soil carbon stabilization through aggregate formation. Elevated CO_2 in the atmosphere has been shown to favor C4 grasses, woody species, and legumes, which all have distinct litter characteristics that in turn influence the microbial communities and carbon dynamics (Pendall et al., 2011). Comparatively, alterations due to climate change occur over shorter timescale in soil communities than the plant communities, which hampers the soil–plant network by decoupling the above-ground and below-ground subsystems (Figure 3.9). The mechanisms for differential responses between above-ground and below-ground organisms to climate

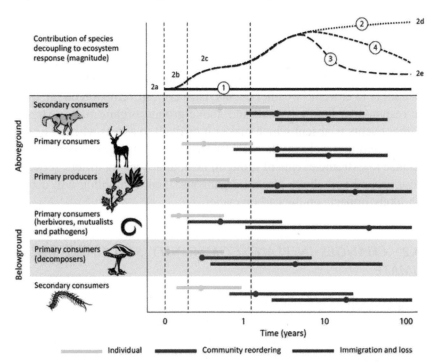

FIGURE 3.9   The hierarchical response model of ecosystem response to climate change. The estimated timescale of response to climate change for each class of mechanism is shown. The top panel shows the hypothetical contribution of species decoupling to ecosystem response under four scenarios: (1) Species remain coupled, or there is no decoupling because species are functionally equivalent; (2) increasing decoupling across time, as outlined in the text; (3) dispersal recouples species interactions system after a period of temporary disequilibria (fairly rapid dampening); and (4) evolution recouples species interactions after a period of temporary disequilibria (slower dampening). (*Source:* Bardgett et al., 2013. Hierarchical Responses of Plant–Soil Interactions to Climate Change: Consequences for the Global Carbon Cycle. *J. Ecol., 101*, 334–343; Used with permission from John Wiley.)

change are unclear, but it may be due to differences in phenology and in the resistance and resilience of above-ground and below-ground communities (Lindberg & Bengtsson, 2006). For example, it has been shown that drought causes immediate death of soil microbes and fauna that pose potentially irreversible consequences for community composition and carbon and nutrient cycling. Moreover, these responses are related to the life history strategies of soil organisms: fast-growing organisms are generally more susceptible to drought, but their populations recover quicker than slow-growing organisms do. Higher trophic levels, which influence nutrient and carbon cycling, also show slow recovery than lower trophic levels. The reason and fate of such differential responses of above-ground and below-ground communities to climate change for carbon cycling are not known. However, the decoupling of networks, both within below-ground food webs and between plant and below-ground communities, is likely to influence future ecosystem resistance and resilience to climate change-related disturbances.

Changes in the abundance of below-ground communities like pathogens and pests can also influence the plant communities and change their response to climate change. An example of this concerns below-ground pathogens that are likely to be affected by climate change phenomena, such as increased growing season length, increased temperature or CO_2, and changes in precipitation.

Increases in temperature can hasten the reproduction rates of below-ground pathogens allowing novel pathogens to thrive. Due to such alteration, sudden disease outbreaks may arise which can affect plant communities and correspondingly ecosystem carbon cycling. Warming might also stimulate horizontal gene transfer between bacterial and fungal species. Such an incidence may generate functionally novel organisms with potential knock-on effects for carbon cycling. The final picture is a highly complex situation where the impact of climate change is distributed over a community structure incorporating plants, grazers, and their predators, which can have significant, but unpredictable, impacts on carbon and nutrient cycling.

3.8.3   CLIMATE AND PLANT SPECIES LOSS

After a considerable period of plant–climate interactions and possible community reordering in various ecosystems, the plant species which fall short in adaptation result in self-omission. As a result of such changes, new individuals arise giving forth new interaction networks operating above and below ground coupled with the discontinuation of the older interactive networks (Wardle et al., 2011). The process of species addition or loss is multifaceted which accounts for significant physiological and ecological alterations. However, recent research shows that climate change has imposed abrupt changes in plant habit and habitat, with some plants dispersing to the poles for existence, tree-line change, upward movement of boreal trees, and so on (Lenoir et al., 2008). Such changes in community reformation may lead to the cessation of existing community functioning which will directly influence the soil environment and processes.

KEYWORDS

- plant life
- soil environment
- soil fertility
- degradation
- soil texture

REFERENCES

Abdul Khalil, H. P. S.; Md. Hossain, S.; Rosamah, E.; Azli, N. A.; Saddon, N.; Davoudpoura, Y.; Md. Islam, N.; Dungani, R. The Role of Soil Properties and Its Interaction towards Quality Plant Fiber: A Review. *Renew. Sustain. Energy Rev.* **2015,** *43,* 1006–1015.

Aiken, R. M.; Smucker, A. J. M. Root System Regulation of Whole Plant Growth. *Ann. Rev. Phytopathol.* **1996,** *34,* 325–346.

Aldaikh, H.; Alexander, N. A.; Ibraim, E.; Oddbjornsson, O. Two Dimensional Numerical and Experimental Models for the Study of Structure–Soil–Structure Interaction Involving Three Buildings. *Comp. Struct.* **2015,** *150,* 79–91.

Aroca, R.; Porcel, R.; Ruiz-Lozano, J. M. Regulation of Root Water Uptake under Abiotic Stress Conditions. *J. Exp. Bot.* **2012,** *63,* 43–57.

Atkin, O. K.; Tjoelker, M. G. Thermal Acclimation and the Dynamic Response of Plant Respiration to Temperature. *Trend Plant Sci.* **2003,** *8,* 343–351.

Bachmair, S.; Weiler, M.; Nutzmann, G. Controls of Land Use and Soil Structure on Water Movement: Lessons for Pollutant Transfer through the Unsaturated Zone. *J. Hydrol.* **2009,** *369,* 241–252.

Bardgett, R. D.; Manning, P.; Morrien, E.; De Vries, F. T. Hierarchical Responses of Plant–Soil Interactions to Climate Change: Consequences for the Global Carbon Cycle. *J. Ecol.* **2013**, *101*, 334–343.

Bardgett, R. D.; Streeter, T. C.; Cole, L.; Hartley, I. R. Linkages between Soil Biota, Nitrogen Availability, and Plant Nitrogen Uptake in a Mountain Ecosystem in the Scottish Highlands. *App. Soil Ecol.* **2002**, *19*, 121–134.

Barre, P.; Velde, B.; Fontaine, C.; Catel, N.; Abbadie, L. Which 2:1 Clay Minerals Are Involved in the Soil Potassium Reservoir? Insights from Potassium Addition or Removal Experiments on Three Temperate Grassland Soil Clay Assemblages. *Geoderma* **2008**, *146*, 216–223.

Bradl, H. B. Absorption of Heavy Metal Ions on Soils and Soil Constituents. *J. Colloids Interface Sci.* **2004**, *277*, 1–18.

Brady, N. C. *The Nature and properties of Soils*, 10th ed.; Prentice Hall of India: New Delhi, 1995.

Brown, S. M.; Petrone, R. M.; Chasmer, L.; Mendoza, C.; Lazerjan, M. S.; Landhausser, S. M.; Silins, U.; Leach, J.; Devito, K. J. Atmosphere and Soil Moisture Controls on Evapotranspiration from above and within a Western Boreal Plain Aspen Forest. *Hydrol. Proc.* **2014**, *28*, 4449–4462.

Clark, M. S.; Horwath, W. R.; Shennan, C.; Scow, K. M. Changes in Soil Chemical Properties Resulting from Organic and Low Input Farming Practices. *Agron. J.* **1998**, *90*, 662–671.

Cotrufo, P.; Ineson, P.; Rowland, A. P. Decomposition of Tree Leaf Litter Grown under Elevated CO_2: Effect of Litter Quality. *Plant Soil* **1994**, *163*, 121–130.

Craine, J. M.; Gelderman, T. M. Soil Moisture Controls on Temperature Sensitivity of Soil Organic Carbon Decomposition for a Mesic Grassland. *Soil Biol. Biochem.* **2011**, *43*, 455–457.

Davies, W. J.; Zhang, J. Root Signals and the Regulation of Growth and Development of Plants in Drying Soil. *Ann. Plant Physiol. Plant Mol. Biol.* **1991**, *42*, 55–76.

Drake, J. E.; Gallet-Budynek, A.; Hofmockel, K. S.; Bernhardt, E. S.; Billings, S. A.; Jackson, R. B.; Johnsen, K. S.; Lichter, J.; McCarthy, H. R.; McCormack, M. L.; Moore, D. J. P.; Oren, R.; Palmroth, S.; Philips, R. P.; Pippen, J. S.; Pritchard, S. G.; Treseder, K. K.; Schlesinger, W. H.; DeLucia, E. H.; Finzi, A. C. Increases in the Flux of Carbon below Ground Stimulate Nitrogen Uptake and Sustain the Long Term Enhancement of Forest Productivity under Elevated CO_2. *Ecol. Lett.* **2011**, *14*, 349–357.

Elmashad, M. E.; Ata, A. A. Effect of Sea Water on Consistency, Infiltration Rate and swelling Characteristics of Montmorillonite Clay. *HBRC J.* **2014**. http://dx.doi.org/10.1016/j.hbrcj.2014.12.004.

Gange, A. C.; Gange, E. G.; Sparks, T. H.; Boddy, L. Rapid and Recent Changes in Fungal Fruiting Patterns. *Science* **2007**, *316*, 71.

Grondona, S. I.; Gonzalez, M.; Martinez, D. E.; Massone, H. E.; Miglioranza, K. S. B. Endosulfan Leaching from Typic Argiudolls in Soyabean Tillage Areas and Ground Water Pollution Implications. *Sci. Tot. Environ.* **2014**, *484*, 146–153.

Harrison-Kirk, T.; Beare, M. H.; Meenken, E. D.; Condron, L. M. Soil Organic Matter and Texture Affect Responses to Dry/Wet Cycles: Changes in Soil Organic Matter Fractions and Relationships with C&N Mineralization. *Soil Biol. Biochem.* **2014**, *74*, 50–60.

Howell, T. A. Irrigation Efficiency. In *Encyclopedia of Water Science*; Stewart, B. A., Howell, T. A., Eds.; Marcel Dekker: New York, 2003, pp 467–472.

Jackson, M. B. Are Plant Hormones Involved in Root to Shoot Communication? *Adv. Bot. Res.* **1993,** *19*, 103–187.

Jackson, R. B.; Cook, C. W.; Pippen, J. S.; Palmer, S. M. Increased Below Ground Biomass and Soil CO_2 Fluxes after Decade of Carbon Di-oxide Enrichment in Warm Temperature Forests. *Ecology* **2009,** *90*, 3352–3366.

Jasechko, S.; Sharp, Z. D.; Gibson, J. J.; Birks, S. J.; Peter, Y. Y.; Fawcett, P. J. Terrestrial Water Fluxes Dominated by Transpiration. *Nature* **2013,** *496*, 347–350.

Jiang, X.; Hou, X.; Zhou, X.; Xin, X.; Wright, A.; Jia, Z. pH Regulates Key Players of Nitrification in Paddy Soils. *Soil Biol. Biochem.* **2015,** *81*, 9–16.

Korner, C.; Assholf, R.; Bignucolo, O.; Hatternschwiler, S.; Keel, S. G.; Pelaez-Riedl, S.; Pepin, S.; Siegwolf, R. T. W.; Zotz, G. Carbon Flux and Growth in Mature Deciduous Trees Exposed to Elevated CO_2. *Science* **2005,** *309*, 1360–1362.

Krumins, J. A.; Goodey, N. M.; Gallagher, F. Plant–Soil Interactions in Metal Contaminated Soils. *Soil Biol. Biochem.* **2015,** *80*, 224–231.

Kukal, S. S.; Sidhu, A. S. Percolation Losses of Water in Relation to Pre-puddling Tillage and Puddling Intensity in a Puddle Sandy Loam Rice (*Oryza sativa* L.) Field. *Soil Till. Res.* **2004,** *78*, 1–8.

Lamsal, S.; Mishra, U. Mapping Soil Textural Fractions across a Large Watershed in North-East Florida. *J. Environ. Manage.* **2010,** *91*, 1686–1694.

Larsen, S.; Widdowson, A. E. Chemical Composition of Soil Solution. *J. Sci. Food Agric.* **1968,** *19*, 693–695.

Lehmann, J.; da Silva Jr., J. P.; Steiner, C.; Nehls, T.; Zech, W.; Glaser, B. Nutrient Availability and Leaching in an Archaeological Anthrosol and a Ferrasol of the Central Amazon Basin: Fertilizer, Manure and Charcoal Amendments. *Plant Soil* **2003,** *249*, 343–357.

Lenoir, J.; Gegout, J. C.; Marquet, P. A.; de Ruffray. P.; Brisse, H. A Significant Upward Shift in Plant Species Optimum Elevation during the 20th Century. *Science* **2008,** *320*, 1768–1771.

Letey, J. The Study of Soil Science: Science or Art. *Aust. J. Soil Res.* **1991,** *29*, 699–707.

Li, S.; Kang, S.; Zhang, L.; Du, T.; Tong, L.; Ding, R.; Gao, W.; Zhao, P.; Chen, X.; Xiao, H. Ecosystem Water Use Efficiency for a Sparse Vineyard in Arid Northwest China. *Agric. Water Manage.* **2015,** *148*, 24–33.

Lindberg, N.; Bengtsson, J. Recovery of Forest Soil Fauna Diversity and Composition under Repeated Summer Droughts. *Oikos* **2006,** *114*, 494–506.

Long, G.-Q.; Jiang, Y.-J.; Sun, B. Seasonal and Inter-annual Variation of Leaching of Dissolved Organic Carbon and Nitrogen under Long Term Manure Application in Acidic Clay Soil in Subtropical China. *Soil Till. Res.* **2015,** *146*, 270–278.

Manalil, S.; Flower, K. Soil Water Conservation and Nitrous Oxide Emissions from Different Crop Sequences and Fallow under Mediterranean Conditions. *Soil Till. Res.* **2014,** *143*, 123–129.

Martin, R. T.; Bailey, S. W.; Eberl, D. D.; Fanning, D. S.; Guggenheim, S.; Kodama, H.; Pevear, D. R.; Srodon, J.; Wicks, F. J. Report of the Clay Minerals Society Nomenclature Committee: Revised Classification of Clay Minerals. *Clay Miner.* **1991,** *39*, 333–335.

McDowell, W. H.; Liptzin, D. Linking Soil and Streams: Response of Soil Solution Chemistry to Simulated Hurricane Disturbance Mirrors Stream Chemistry Following a Severe Hurricane. *For. Ecol. Manage.* **2014**, *332*, 56–63.

Mendoza, R. E.; Garcia, I. V.; de Cabo, L.; Weigandt, C. F.; de Iorio, A. F. The Interaction of Heavy Metals and Nutrients Present in Soil and Native Plants with *Arbuscular mycorrhizae* on the Riverside in Matanza-Riachuelo River Basin (Argentina). *Sci. Tot. Environ.* **2014**, *505*, 555–564.

Misra, A.; Tyler, G. Influence of Soil Moisture on Soil Solution Chemistry and Concentration of Minerals in the Calcioles *Phleum phleoides* and *Veronica spicata* Grown on a Limestone Soil. *Ann. Bot.* **1999**, *84*, 401–410.

Muilu-Makela, R.; Vuosku, J.; Laara, E.; Saarinen, M.; Heiskanen, J.; Haggman, H.; Sarjala, T. Water Availability Influences Morphology, Mycorrhizal Associations, PS II Efficiency and Polyamine Metabolism at Early Stage Growth Phase of Scots Pine Seedlings. *Plant Physiol. Biochem.* **2015**, *88*, 70–81.

Nguyen, M. N.; Dultz, S.; Tran, T. T. T.; Bui, A. T. K. Effect of Anions on Dispersion of a Kaolinitic Soil Clay: A Combined Study of Dynamic Light Scattering and Test Tube Experiments. *Geoderma* **2013**, *209–210*, 209–213.

Odhiambo, L. O.; Irmak, S. Evaluation of the Impact of Surface Residue Cover on Single and Dual Crop Co-efficient for Estimating Soyabean Actual Evapotranspiration. *Agric. Water Manage.* **2012**, *104*, 221–234.

Pailler, A.; Vennetier, M.; Torre, F.; Ripert, C.; Guiral, D. Forest Soil Microbial Functional Patterns and Response to a Drought and Warming Event: Key Role of Climate–Plant–Soil Interactions at a Regional Scale. *Soil Biol. Biochem.* **2014**, *70*, 1–4.

Passioura, J. B. Soil Conditions and Plant Growth. *Plant Cell Environ.* **2002**, *25*, 311–318.

Pendall, E.; Osanai, Y.; Williams, A. L.; Hovenden, M. J. Soil Carbon Storage under Simulated Climate is Mediated by Plant Functional Type. *Glob. Change Biol.* **2011**, *17*, 505–514.

Peng, X.; Horn, R.; Hallett, P. Soil Structure and its Function in Ecosystems: Phase Matter and Scale Matter. *Soil Till. Res.* **2015**, *146*, 1–3.

Pypers, P.; Sanginga, J.-M.; Kaserekra, B.; Walangululu, M.; Vanlauwe, B. Increased Productivity through Integrated Soil Fertility Management in Cassava–Legume Intercropping Systems in the Highlands of Sud-Kiva, DR Congo. *Field Crop Res.* **2011**, *120*, 76–85.

Quan, S.-X.; Yan, B.; Lei, C.; Yang, F.; Li, N.; Xiao, X.-M.; Fu, J.-M. Distribution of Heavy Metal Pollution in Sediments from an Acid Leaching Site of E-waste. *Sci. Tot. Environ.* **2014**, *499*, 349–355.

Rahil, M. H.; Qanadillo, A. Effect of Different Irrigation Regimes on Yield and Water Use Efficiency of Cucumber Crop. *Agric. Water Manage.* **2015**, *148*, 10–15.

Rajaie, M.; Karimian, N.; Yasrebi, J. Nickel Transformation in Two Calcareous Soil Textural Classes as Affected by Applied Nickel Sulfate. *Geoderma* **2008**, *144*, 344–351.

Reddy, Y. T.; Reddy, S. G. H. *Principles of Agronomy*, 4th ed.; Kalyani Publishers: New Delhi, 2010.

Sanchez, P. A.; Leaky, R. R. B. Landuse Transformation in Africa: Three Determinants for Balancing Food Security with Natural Resource Utilization. *Eur. J. Agron.* **1997**, *7*, 1–9.

Sheil, D. How Plants Water Our Planet: Advances and Imperatives. *Trends Plant Sci.* **2014**, *19*, 209–211.

Shukla, S.; Shrestha, N. K.; Jaber, F. H.; Srivastava, S.; Obreza, T. A.; Boman, B. J. Evapotranspiration and Crop Co-efficient for Water Melon Grown under Plastic Mulched Conditions in Sub-tropical Florida. *Agric. Water Manage.* **2014,** *132,* 1–9.

Smith, M. D.; Knapp, A. K.; Collins, S. L. A Framework for Assessing Ecosystem Dynamics in Response to Chronic Resource Alterations Induced by Global Change. *Ecology* **2009,** *90,* 3279–3289.

Stark, H.; Nothdurft, A.; Block, J.; Bauhus, J. Forest Restoration with *Betula* spp. and *Populus* spp. Nurse Crops Increases Productivity and Soil Fertility. *For. Ecol. Manage.* **2015,** *339,* 57–70.

Tokumoto, I.; Noborio, K.; Koga, K. Coupled Water and Heat Flow in a Grass Field with Aggregated Andisol during Soil Freezing Periods. *Cold Reg. Sci. Technol.* **2010,** *62,* 98–106.

Vaidyanathan, L. V. Influence of Organic Carbon and Nitrogen Associated with Cropping History and Soil Textural Class on Nitrogen Use by Winter Wheat (*Triticum aestivum*). *Dev. Agric. Manage. For. Ecol.* **1992,** *25,* 57–68.

Verburg, K.; Bond, W. J.; Hunt, J. R. Fallow Management in Dryland Agriculture: Explaining Soil Water Accumulation using a Pulse Paradigm. *Field Crop Res.* **2012,** *130,* 68–79.

Vicente-Serrano, S. M.; Van der Schier, G.; Begueria, S.; Azorin-Molina, C.; Lopez-Moreno, J.-I. Contribution of Precipitation and Reference Evapotranspiration to Drought Indices under Different Climates. *J. Hydrol.* **2014.** http://dx.doi.org/10.1016/j.jhydrol.2104.11.025.

Wang, E.; Cruse, R. M.; Zhao, Y.; Chen, X. Quantifying Soil Physical Condition Based on Soil Solid, Liquid and Gaseous Phases. *Soil Tillage Res.* **2015,** *146,* 4–9.

Wang, P.; Song, X.; Han, D.; Zhang, Y.; Zhang, B. Determination of Evaporation, Transpiration and Deep Percolation of Summer Corn and Winter Wheat after Irrigation. *Agric. Water Manage.* **2012,** *105,* 32–37.

Wang, Q.; Cheng, T.; Wu, Y. Influence of Mineral Colloids and Humic Substances on Uranium (VI) Transport in Water-Saturated Geologic Porous Media. *J. Cont. Hydrol.* **2014,** *170,* 76–85.

Wardle, D. A.; Bardgett, R. D.; Callaway, R. M.; Van der Putten, W. H. Terrestrial Ecosystem Responses to Species Gains and Losses. *Science* **2011,** *332,* 1273–1277.

Wu, Y. Z.; Huang, M. B.; Warrington, D. N. Black Locust Transpiration Responses to Soil Water Availability as Affected by Meteorological Factors and Soil Texture. *Pedosphere* **2015,** *25,* 57–71.

Wu, Z.; Dijkstra, P.; Koch, G. W.; Penuelas, J.; Hungate, B. A. Responses of Terrestrial Ecosystems to Temperature and Precipitation Change: A Meta-analysis of Experimental Manipulation. *Global Change Biol.* **2011,** *17,* 927–942.

Xiao, J.; Sun, G.; Chen, J.; Chen, H.; Chen, S.; Dong, G.; Kato, T. Carbon Fluxes, Evapotranspiration, and Water Use Efficiency of Terrestrial Ecosystems in China. *Agric. For. Meteorol.* **2013,** *182,* 76–90.

Yuste, J. C.; Baldocchi, D. D.; Gershenson, A.; Goldstein, A.; Misson, L.; Wong, S. Microbial Soil Respiration and its Dependency on Carbon Inputs, Soil Temperature and Moisture. *Glob. Change Biol.* **2007,** *13,* 2018–2035.

Zha, T.; Li, C.; Kellomaki, S.; Peltola, H.; Wang, K.-Y.; Zhang, Y. Control of Evapotranspiration and CO_2 Fluxes from Scots Pine by Surface Conductance and Abiotic Factors. *PLoS ONE* **2013.** http://dx.doi.org/10.1371/journal.pone.0069027.

Zhang, K.-X.; Pan, S.-M.; Zhang, W.; Xu, Y.-H.; Cao, L.-G.; Hao, Y.-P.; Wang, Y. Influence of Climate on Reference Evapotranspiration and Aridity Index and Their Temporal-Spatial Variations in the Yellow River Basin, China, from 1961–2012. *Quat. Int.* **2015**. http://dx.doi.org/10.1016/j.quaint.2014.12.037.

Zhongjie, S.; Yanhui, W.; Pengtao, Y.; Lihong, X.; Wei, X.; Hao, G. Effect of Rock Fragments on the Percolation and Evaporation of Forest Soil in Liupan Mountains, China. *Acta Ecol. Sin.* **2008**, *28*, 6090–6098.

Zirkler, D.; Lang, F.; Kaupenjohann. "Loss in filtration"—The Separation of Soil Colloids from Larger Particles. *Colloids Surf. A: Physiochem. Eng. Asp.* **2012**, *399*, 35–40.

CHAPTER 4

PLANT ESSENTIAL NUTRIENTS, NUTRIENT COMPOSITION IN SOIL, AND ACTIVITY OF PLANT METABOLITES

SUBHASISH DAS and SATYA SUNDAR BHATTACHARYA*

Department of Environmental Science, Tezpur University, Tezpur 784028, Assam, India

Corresponding author, Tel.: +91 3712 267007/+91 3712 2670078/+91 3712 2670079x5610; E-mail: satyasundarb@yahoo. co.in; satya72@tezu.ernet.in.

CONTENTS

ABSTRACT

Plant nutrients rely on the composition of the soil. With the improper health of the soil, the condition of plants gets devastated. Therefore, nutrition of plants truly is affected by the nutrient of soil. The transport of nutrients from soil to the plants, plants to soil, and plants to climate depends mostly on the type and habit of plants. All these intricate relations between the functional compartments of the soil–plant–climate continuum highly influence the overall development of any ecosystem.

4.1   PLANT AND SOIL

It is well known that roots of terrestrial plants acquire nourishment from the soil. The nourishment is obtained in the form of certain nutrient elements referred to as essential elements (Tisdale et al., 1995). However, most of the agricultural soils around the globe present a significant variability in their nutrient budget and also record a prominent downfall in the availability of essential plant nutrients (Yildirin et al., 2011). Roots, principally as inorganic ions, absorb these elements. These ions in soils are derived mostly from mineral constituents of the soil. A mineral nutrient can be referred to an inorganic ion obtained from the soil and required for plant growth. Poor nutrient holding of a typical soil is substituted with externally applied fertilizers. These mineral fertilizers are applied either by direct administration into the soil system or as foliar spraying. However, research shows that the chemical fertilizers are less efficient in supplying plants with their requisite level of mineral nutrition (Adesemoye et al., 2009). Mineral nutrition can be defined as a cumulative process encompassing absorption, translocation, and assimilation of nutrients by plants. A variation in vegetation type or alteration in the same may result in marked hindrance in nutrient biogeochemical cycling. It is also noteworthy that such plants resist the loss of nutrients from the ecosystem (Boligar et al., 2001). A complex interaction operates between plants, soil, and climate which generates feedbacks that in turn regulate the nutrient uptake efficiency of plants. All these intricate relations between the functional compartments of the soil–plant–climate continuum highly influence the overall development of any ecosystem (Argen et al., 2012). Plants are source of various chemical moieties, which are produced because of their primary physiological processes or different secondary metabolism. The primary metabolites like plant pigments (e.g., chlorophyll, caroteinoid, xanthophylls, etc.) aid in photosynthesis, act as substrates for different physiological pathways, impart color to the organism, and so on. However, the secondary metabolites like phenols, flavonoids, terpenenoids, alkaloids, and so on, are produced during stress conditions or as a compound for defense to resist herbivory or exclusion. The polyphenolic compounds possess one (phenol) or more (polyphenol) hydroxyl substituent along with other functional groups such as esters, methyl esters, and glycosides (Harborne, 1989). These metabolites represent approximately 40% of organic carbon circulating in the biosphere. Plants require such compounds for growth, reproduction, pigmentation, and disease resistance. These polymeric structures are localized in several plant parts like leaves, flowers, roots, or whole body itself. For example, about 8150 or more are flavonoids and found in the epidermis of leaves and fruit skin of various plants (Lattenzio et al., 2006).

4.2 PLANT NUTRIENTS AND THE CRITERIA OF ESSENTIALITY

A mineral nutrient is considered essential to plant growth and development if the element is involved in plant metabolic functions and the plant cannot complete its life cycle without the element (Epstein, 1994). Usually, the plant shows an observable symptom indicating deficiency in a specific nutrient, which can be rectified by supplying the same. The following terms are commonly used to designate the levels of nutrients in plants (Tisdale et al., 1995).

Deficient: If the concentration of an essential element is severely low that might limit the crop yield and distinguishable symptoms are visible on the plant body then the plant can be called nutrient deficient. In case of extreme deficiency, plant death is inevitable. With slight or moderate deficiency, the yield of the plant will be reduced.

Critical range: It is the threshold of nutrient concentration in the plant below which the yield response to added nutrient occurs. The critical range usually varies among plants and nutrients, but occurs somewhere in the transition between nutrient deficiency and sufficiency.

Sufficient: It is the range of nutrient concentration wherein added nutrient will not increase yield but can increase the overall nutrient budget. The term "luxury consumption" is generally used to describe the mode of nutrient absorption by plants that does not influence yield but increases the plant biomass.

Excessive: If the concentration of essential or other elements is high enough to reduce plant growth and yield, then it is termed as nutrient excessive. Excessive nutrient concentration can cause an imbalance in other essential nutrients, which indirectly reduces yield.

Soil nutrient concentration and plant yield are directly proportional to each other. Yield of the crop gets severely affected if a nutrient-deficient situation is not rectified (Velarde et al., 2005). Under severe deficiency conditions, high crop yield with externally added nutrient can cause a condition of meager nutrient deficiency. This particular phenomenon is referred to as the Steenberg effect which arises due to the dilution of the nutrient in the plant body by its swift growth (Fig. 4.1). When the concentration reaches the critical range, plant yield is generally maximized. However, an increase in the nutrient concentration beyond critical range implies that the plant is taking up nutrient above its requirement which again may reduce its yield directly through toxicity or indirectly by lowering concentrations of other nutrients below their critical ranges (Tisdale et al., 1995).

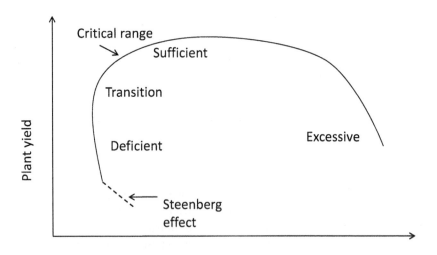

FIGURE 4.1   Diagrammatic representation of Steenberg effect.

Seventeen elements are considered essential to plant growth; their average concentrations in plants are listed in Table 4.1. Carbon (C), hydrogen (H), and oxygen (O) are the most abundant elements in plants. These elements are essential for running the major physiological processes, like production of amino acids, sugars, proteins, nucleic acids, and others, and are termed as non-mineral nutrients (Koorem et al., 2014). The rest 14 essential elements are the mineral nutrients, classified as macronutrients and micronutrients. The basis of classification is their relative abundance in plants. The macronutrients are nitrogen (N), phosphorus (P), potassium (K), sulfur (S), calcium (Ca), and magnesium (Mg), whereas the micronutrients are iron (Fe), zinc (Zn), manganese (Mn), copper (Cu), boron (B), chlorine (Cl), nickel (Ni), and molybdenum (Mo). Apart from these, there are four other elements like sodium (Na), cobalt (Co), vanadium (Va), and silicon (Si) under the hood of essential micronutrients in some plants. A deficiency in micronutrients can also reduce plant yield like macronutrient deficiency (Velarde et al., 2005; Heidak et al., 2014; Brown et al., 1987).

TABLE 4.1   Average Plant Nutrient.

Plant nutrient	Average concentration (on dry weight basis) (%)	Reference
H	6	Tisdale et al. (1995)
O	45	
C	45	
N	1.5	
K	1	
Ca	0.5	
Mg	0.2	
P	0.2	
S	0.1	
Cl	0.01	
Fe	0.01	
B	0.002	
Mn	0.005	
Zn	0.002	
Cu	0.0006	
Mo	0.00001	

Source: Tisdale, S. L.; Nelson, W. L.; Beaton, J. D. and Havlin, J. L., Soil Fertility and Fertilizers, 5th ed., Prentice Hall of India, New Delhi, 1995.

When plant material is burned, the resultant ash residue contains all essential and nonessential mineral elements other than C, H, O, N, and S, which flee off as gases (Rossi et al., 2013). The amount of nutrients in the plants is affected by various factors making their respective concentrations vary significantly from each other (shown in Table 4.2). Knowledge on plant nutrient concentration levels is important for ascertaining the nutrient requirement for certain plants and thus can be useful for establishing fertilizer dose recommendations (Muhammad et al., 2015). Due to the dynamic soil reactions, both biologically and chemically, the amount of nutrients absorbed by plants does not equal the amount applied as fertilizer. The nutrients applied to the plants for their growth and development finally return back into the soil after its senescence. It is after their death, that the complex forms of the nutrients are broken down by biochemical reactions in soil and added into the overall nutrient budget of the soil (Da Silva Mastos et al., 2011; Partey et al., 2011).

TABLE 4.2   Approximate Utilization of Nutrients by Selected Crops.

Plant	Yield (acre)	N (in lb)	P (in lb)	K (in lb)	Mg (in lb)	S (in lb)	Reference
Alfalfa	10 t	600	52	500	53	51	Tisdale et al. (1995)
Orchard grass	6 t	300	44	313	25	35	
Coastal Bermuda	10 t	500	61	350	50	50	
Clover grass	6 t	300	39	300	50	50	
Corn							
Grain	200 bu	150	40	48	18	15	
Stover	8000 lb	116	12	174	47	18	
Sorgum							
Grain	8000 lb	120	26	25	14	22	
Stover	8000 lb	130	13	142	30	16	
Corn silage	32 t	266	50	222	65	33	
Cotton							
1500 lb lint, 2250 lb seed		94	17	37	11	7	
Stalks, leaves and burrs		86	11	68	24	23	
Oats							
Grain		80	11	17	5	8	
Straw	100 bu	35	7	104	15	11	

TABLE 4.2   *(Continued)*

Peanuts						
Nuts	4000 lb	140	10	29	5	10
Vines	5000 lb	100	7	125	20	11
Potatoes, Irish						
Tubers	500 cwt	173	32	233	14	15
Vines	300 cwt	96	7	221	36	7
Potatoes, Sweet						
Roots		73	15	140	8	–
Vines	300 cwt	83	13	120	10	–
Rice						
Grain	7000 lb	77	20	23	8	5
Straw	7000 lb	35	6	100	6	7
Soybeans						
Grain	60 bu	240	20	70	17	12
Straw	7000 lb	84	7	48	10	13
Tomatoes						
Fruit	30 t	100	10	180	8	21
Vines	4400 lb	80	11	100	20	20
Wheat						
Grain	80 bu	92	19	23	12	5
Straw	6000 lb	42	4	113	12	15
Barley						
Grain		110	17	29	8	10
Straw	100 bu	40	7	96	9	10
Sugar beets						
Roots	30 t	125	7	208	27	10
Tops	16 t	130	11	250	53	35
Sugarcane						
Stalks		160	39	279	40	54
Tops and trash	100 t	200	29	229	60	32

Source: Tisdale, S. L.; Nelson, W. L.; Beaton, J. D. and Havlin, J. L., Soil Fertility and Fertilizers, 5th ed., Prentice Hall of India, New Delhi, 1995.

4.3 BIOAVAILABILITY OF NUTRIENTS

Bioavailability of nutrients pertains to the mechanism by which plant roots can absorb the essential elements from the soil solution (Silver, 1994). The capacity of plants to retort efficiently to nutrient availability is of paramount importance for their adaptation to the environmental condition (Lope-Bucio et al., 2003). The nutrient supply to plant roots is a very dynamic process. Plants take up cations and anions from the soil solution and in exchange give away small quantities of ions such as H^+, OH^-, and HCO_3^- into the soil solution. Changes in ion concentrations in soil solution are "buffered" by ions adsorbed on surfaces of soil minerals. The removal of the same ions from the soil solutions gives rise to partial desorption of the same ions from mineral surfaces. Soil abounds in mineral compounds that have the ability to dissolve and resupply many ions back into the soil solution. Similarly, an increase in the ion concentration in soil solution may cause a saturated state leading to precipitation of some minerals (Tunesi et al., 1999).

There are several factors that influence the bioavailability of nutrients (Comerford, 2005):

- Nutrient supply capacity of the soil which is governed by its parent material and hosted vegetation,
- Soil pH as it moderates the nutrient traffic,
- Microbial activity in the soil that initiates the decay and nutrient release processes,
- External addition of nutrient sources like commercial fertilizer, manure, etc.
- Soil physical environment attributes like temperature, moisture, etc., and
- Vegetation covers of the soil that modifies the soil environment.

4.3.1 TRANSPORT OF NUTRIENTS TO ROOT SURFACE

The release and uptake of nutrients is a complex process. Transport of nutrients from soil to the plant root tissue and eventually to the aerial parts of the plants is a key mechanism which must work efficiently for healthy plant growth (Tester & Leigh, 2001). The nutrients remain in dynamic equilibria before getting absorbed by the plant roots. Root-nutrient acquisition involves various processes like proliferation, transport, exudation, symbioses, and the delivery of dissolved nutrients from the soil solution to the root surface by mass flow and diffusion (Chapman et al., 2012). However, there are two theories that describe the mechanism of nutrient uptake by plants.

Soil solution theory: The nutrients in soil are dissolved in the soil solution which flow to root surfaces following the principles of mass flow and diffusion. Mass flow can be popularly designated as the movement of ions along with moving water to the roots. The nutrient movement relies on the rate of water flow. Diffusion takes place when a concentration gradient occurs between soil solution and roots leading to the flow of nutrients to the roots that are maintained at low-nutrient gradient.

Contact exchange theory: This theory says that due to the close contact between root surfaces and soil colloids, a direct exchange of H^+ from the plant roots occurs in exchange for cations from soil colloid. The contact exchange theory seems less viable than the soil solution theory when it comes to justification of the nutrient transport mechanism to and fro the plant–soil systems.

The mechanism of nutrient transport is a dynamic process. The rate of nutrient uptake depends mostly on its concentration in soil solution. As soon as the roots absorb nutrients, its deficiency occurs in the rhizosphere that is eventually replenished by mass flow or diffusion of nutrients. The capacity of the soil to manage sufficient nutrient level is vital. The following scheme illustrates a common nutrient transport for most of the important nutrient elements:

Unavailable $\xleftrightarrow{(a)}$ Intermediate $\xleftrightarrow{(b)}$ Labile $\xleftrightarrow{(c)}$ Nutrient in soil solution

The constants (a), (b), and (c) are rate constants. The equilibrium between unavailable and intermediate form is established slowly. The intermediate forms are the long-term reserves that can be replenished slowly from inert forms or more rapidly by fertilizer reactions with soil minerals. The equilibrium between intermediate and labile forms is established over a short period. The labile nutrient is loosely held and is a fraction of a soil nutrient that comes to equilibrium with soil solution rapidly. Generally, the labile pool represents the main segment of the quantity factor while the nutrient concentration of the soil solution is the intensity factor. Nutrient absorption by plant roots is directly dependent on the concentration of the soil solution (intensity factor) which in turn is regulated by the labile pool (quantity factor).

Environment plays a major role in devising and regulating the nutrient acquisition phenomenon of plants (Lemoine et al., 2013). There are several other environmental attributes which influence the mechanism of nutrient absorption by plants. They can be seen as follows:

External factors

- oxidation–reduction state of elements
- concentration of the elements
- moisture content of soil
- aeration
- temperature
- pH.

Internal factors

- cell wall
- aeration
- type of cells and stage of development
- transpiration
- rhizosphere.

4.4 NUTRIENT TRANSFORMATION IN SOIL

The nutrients present in soil are subjected to physical, chemical, and biological changes which transform these nutrients into the bioavailable form for easy plant uptake (Sharma et al., 1992). The nutrient transformation is governed by various biotic and abiotic factors. During these transformations, a series of intermediate products are formed which may or may not be beneficial to plants. These transformations may either release or fix nutrients. Some of the major nutrient transformations in soil are discussed below.

4.4.1 NITROGEN TRANSFORMATION

Nitrogen is an important macroelement which undergoes major microbially aided transformations before plant uptake (Robertson & Groffman, 2015). After getting mineralized of organic matter, ammonium and nitrate are released that are plant available. Nitrate may be further transformed into molecular nitrogen and escape into atmosphere. Mineralization is highly dependent on temperature and water retention capacity of the soil. Thus, higher the temperature and water retention capacity, higher will be the mineralization. Fertilizers added to the soils are transformed and several losses occur during this process. Immediately, after application of urea, it is hydrolyzed in the presence of enzyme urease and forms ammonium carbonate. Ammonium carbonate is an unstable compound that eventually decomposes into ammonium and carbon dioxide. Ammonium is adsorbed on the clay complex, a portion is absorbed by the crop and most of it is lost as volatile ammonia gas. The several forms of applied and native N are prone to various losses, like volatilization, leaching, denitrification, ammonium fixation, and immobilization (Povilaitis et al., 2014).

Volatilization: The alteration of material from solid to vapor form is called volatilization. Urea and ammoniacal fertilizers are subjected to volatilization. Urea hydrolyzes and ammonium carbonate is formed which dissociates and increases the pH around urea granules. The increase in pH initiates volatilization losses (Vandre & Clemens, 1997).

Leaching: Loss of nutrients beyond root zone along with water is known as leaching. Soluble fertilizers move along with water into deeper layers. Comparatively, nitrate forms of N fertilizers are more prone to leaching than their ammoniacal counterparts. The ammoniacal fertilizers after dissociation deposit the ammonium ions on the clay surfaces. Scott et al. (2015) have recently reported that variation in soil P content may have a strong influence of N leaching dynamics of soil. Addition of biogas residue in the soil has a positive impact on N leaching from soil (Svoboda et al., 2013).

Nitrogen fixation: It is a primary process which leads to the fixation of atmospheric N into the soil by symbiotic bacteria or free living bacteria in the soil system (Fig. 4.2). This mechanism is aided by an enzyme nitrogenase. The enzyme nitrogenase consists of two components called Protein 1 (molybdoferredoxin) and Protein 2 (azoferredoxin). Azoferredoxin is an oxygen-sensitive iron–sulfur complex whereas molybdoferredoxin is comprised of two different peptide chains forming a tetramer. The mechanism of nitrogenase action is outlined in the following figure (Goodwin & Mercer, 1998).

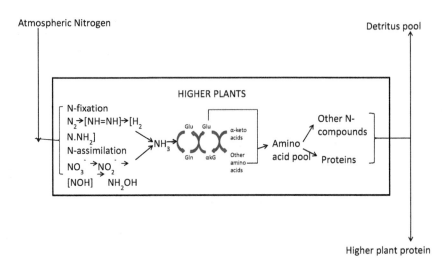

FIGURE 4.2   A portion of nitrogen transformation inside plant tissue (schematic diagram).

Nitrification: The process of conversion of fixed NH_4^+ into NO_3^- is called nitrification. Two main bacteria are associated as key drivers of this process: *Nitrosomonas* and *Nitrobacter*. The representative equations describing the action of these two bacterial strains are provided below. The internal transformation of NO_3^- inside the plant tissue is mediated by the enzyme nitrate reductase (refer to Section 4.6.1).

$$NH_4^+ + 1.5O \longrightarrow NO_2^- + 2H^+ + H_2O$$

$$\textit{Nitrobacter}$$
$$NO_2^- + 0.5O \longrightarrow NO_3^-$$

Denitrification: This process is aided by the denitrifying bacteria present in the soil. Nitrates present in the soil are converted into elemental N and are lost into the atmosphere. Denitrification losses occur more in case of submerged soil (Estavillo et al., 1996).

Ammonium fixation: Fixation of ammonium is due to trapping of these ions within the crystal lattice of montmorillonite, illite, and vermiculite minerals. They fix more ammonium when the soil is dry due to contraction of these minerals. After expansion, NH_4^+ is released with an exchange of Ca^{2+}, Mg^{2+}, Na^+, and H^+ ions. Fixed ammonium is available to microorganisms which convert them to nitrate and also to some plants which absorb those (Nieder et al., 2011).

Immobilization: Immobilization is the temporary locking up of N in microorganisms. It is the opposite of mineralization that deals mostly with the mobilization of nutrients from their complex forms. This process is carried out by fungi, actinomycetes, and bacteria. Mineralization is faster in aerobic soil. When large quantities of organic matter are added to the soil, the microorganisms utilize available N in the soil for their multiplication. N immobilization in soil is affected by a plethora of processes like liming, acidification, N-fertilizer application, straw incorporation, redox potential, and so on (Said-Pullicino et al., 2014; Bergholm et al., 2015).

4.4.2 PHOSPHOROUS TRANSFORMATION

Phosphorous is present in soil both in organic and inorganic forms but inorganic P is comparatively more than its organic form. Mineralization of organic matter releases P in available form (Sanyal & De Datta, 1991). The major forms of inorganic P in soil are Ca–P, Fe–P, and Al–P. P in soil solution is available in the form of primary and secondary orthophosphates ($H_2PO_4^-$ and HPO_4^{2-}). At low pH, availability of primary orthophosphates is more while secondary orthophosphates are available more toward an alkaline range. P in soil solution depends on rate of decomposition of organic matter and rate of reaction with inorganic fraction. Organic and inorganic P persists in equilibrium with P in soil solution. Addition of fertilizers or rapid mineralization processe temporarily changes the P equilibrium in soil. When fertilizers containing soluble phosphates are added to the soil, P is retained or fixed by different mechanisms. Fixation of P is by adsorption, isomorphous substitution, and double decomposition (Tate, 1984). Insoluble phosphates are weathered and released into the soil solution. P fixation is influenced by several factors like pH, nature and amount of clay, free oxides of Fe and Al, $CaCO_3$, and organic matter. At pH 2–5, fixation is mainly by formation of insoluble Fe and Al phosphates. Within pH 6–10, phosphate is precipitated as phosphates of Ca or Mg. Vermiculite and smectite clay fix more phosphates than kaolinite. Aluminum oxides are more important in P fixation than Fe oxides. Phosphate fixation is high in soils with $CaCO_3$. Soluble phosphates are first converted into tricalcium phosphates and subsequently to carbonate apatite.

Plants play a crucial role in the chemical transformation of P in soil (Rosolem et al., 2014). Some of the processes are hydroxylation of soil colloids, chelation of metal ions, or change of the desorption–adsorption processes. Due to such ongoing processes, soil and fertilizer P is continuously withdrawn and absorbed by plants. P is comparatively less mobile than N and application of farm yard manure, phosphorus-rich compost, phosphorus-enriched organic manure, and so on, shortens the period of P fixation and promotes its availability (Omar, 2014; Manjunthaiah, 2003; Helal et al., 2013). P sorption characteristics vary greatly and accordingly significant differences are observed in the P-supplying capacity and P adsorption of the soils. In long-term fertilizer trails, it is seen that total P increases when P fertilizers are applied. Organic bound P is relatively more mobile than inorganic fertilizer P. Mineralization of P is more influenced by soil moisture than temperature. Depending on the type of crop, P acquisition capacity differs. For example, maize can acquire thrice the amount of soil P than ground nut. Other than this, plant phosphate uptake is significantly enhanced by the presence of carboxylates in the rhizosphere.

4.4.3 POTASSIUM TRANSFORMATION

K is an important macronutrient that plays a crucial role in plant metabolism, and it undergoes several transformations before uptake by plants (Meena & Biswas, 2014). Water-soluble K is present in the soil solution as K^+ ion. It is in equilibrium with exchangeable K. Exchangeable K is present only on clay minerals which is adsorbed. Fixed K is bound between the units of clay minerals. The three different forms of K are in dynamic equilibrium with each other.

Fixed K ⟷ Exchangeable K ⟷ Water-soluble K

In case of depletion of one form, the other available forms substitute the loss. Due to weathering of K-rich rocks like mica and feldspar, the unavailable or fixed K are made available for plant uptake. Exchangeable K adsorbed on clay complex is released into soil solution when its concentration is decreased due to plant absorption. When fertilizer is added, K in the soil solution is increased. A fraction of it goes to exchangeable form and in due course even to fixed form.

4.4.4   CALCIUM TRANSFORMATION

Calcium enters soil and gets available to plants after the weathering of rocks like dolomite, calcite, apatite, and calcium feldspar. It may be taken up by plants, lost in drainage, reabsorbed by clay or a small fraction may be reprecipitated as secondary Ca compound (Sato et al., 2009). Ca is not fixed or made unavailable in the soil. Ca remains in dynamic equilibrium between the exchangeable surfaces and soil solution. Higher concentration of Ca in solution is adsorbed on clay minerals. Ca is released into soil solution by diffusion when its concentration is low in soil.

4.4.5   MAGNESIUM TRANSFORMATION

Mg is available to plants due to weathering of biotite, dolomite, chlorate, serpentine, palygorskite, and olivine (Salehi & Tahamtani, 2012). Moreover, the cropping pattern also influences the Mg dynamics in soil. The released Mg may be absorbed by plants, lost in drainage, or reprecipitated as secondary mineral. Mg in soil solution and exchangeable form are in dynamic equilibrium, similar to Ca.

4.4.6   SULFUR TRANSFORMATION

Sulfur occurs in soil as sulfides, sulfates, and in organic combination with C and N. Sulfur traces its origin from the metal sulfides of plutonic rocks. After weathering, these sulfides are oxidized to sulfates. The nature of applied S is in the form of sulfates of Ca, Mg, K, Na, or NH_4 in soil solution in arable soils. These sulfates may be adsorbed as 1:1 clay or hydrous oxides of Fe and Al. It may be absorbed by plants. Soil type is an important attribute which governs S transformation in soil (Vong et al., 2007). Sulfates are reduced to sulfides in water-logged soils and form H_2S, FeS, and others. Soil microorganisms mainly oxidize the elemental S to sulfates.

4.4.7   IRON TRANSFORMATION

Of all the metal elements in the earth's crust, Fe is the most abundant. It occurs in soil as oxides, hydroxides, and phosphates in primary and secondary minerals. A small amount of Fe is released due to weathering of primary and secondary minerals. Fe deficiency occurs in calcareous soils and soils with high P. The solubility of Fe decreases with an increase in pH making it unavailable to the plants (Ming et al., 2011). Higher concentration of P causes deposition of Fe on or inside the root surfaces, possibly in legumes. Oxidation reduces the availability of Fe. In case of submerged soil, Fe availability increases due to reduction of insoluble ferric compounds to soluble ferrous compounds. Organic matter, native or added, improves the availability of Fe.

4.4.8 MANGANESE TRANSFORMATION

Mn-rich minerals (primary minerals) are present mainly in the form of oxides, carbonates, and silicates. Pyrolusite and manganite are the important secondary minerals. Mn in soil mainly exists in Mn^{2+} and MnO_2 forms. At low pH, availability of Mn increases. Its availability is more in submerged soils due to reduction of insoluble manganic compounds to soluble manganous compounds. At high pH and oxidizing conditions, the tri- and tetravalent forms of Mn are available. $CaCO_3$ decreases Mn availability due to strong binding of Mn^{2+} on $CaCO_3$. Addition of organic matter in soil decreases the Mn availability (Mandal & Mitra, 1982; Patrick et al., 1968).

4.4.9 COPPER TRANSFORMATION

Chalcopyrite is the most abundant mineral of Cu in earth's crust. It is also present in secondary minerals and organic complex. Weathering process releases Cu and Cu^{2+} is the plant available form. The solubility of applied Cu fertilizer decreases with a rise in soil pH. It is due to adsorption of Cu^{2+} on soil colloids and due to precipitation as Cu-hydroxides. Cu is fixed strongly on organic matter. The diffusion of Cu in the soil–water system may also be influenced by several anthropogenic activities (Balint et al., 2015).

4.4.10 ZINC TRANSFORMATION

Zinc in earth's crust is found chiefly as sphalerite, a sulfide mineral. Zn also occurs as constituent of organic matter. When soluble Zn-salt is added to the soil, it undergoes several transformations by adsorption, precipitation, entry into clay lattice, and by microbial consumption. Zn transformation in soil is highly influenced by pH. Zn adsorption by clay and organic matter is high at high pH, making Zn only available in acidic conditions. Most of the Zn in soil is present in residual fraction which is considered to be silicates. The different forms of Zn in soil are water soluble, exchangeable, organic, sesquioxide. Incorporation of Zn either in organic or inorganic forms influences the Zn transformation and availability in soil (Xiang et al., 1995). Initially, Zn is reactive in state but then Zn is transformed into unreactive forms subsequently and is associated with Fe and Mn oxides. The transformation of Zn to unreactive form is enhanced by high temperature and also by swelling and shrinkage of soil. Of the various fractions, water-soluble plus exchangeable Zn, amorphous oxides, and carbonate-bound Zn, and Zn-bound crystalline oxides increase, whereas the complexed Zn, organically bound Zn and residual Zn decease with increasing water sodicity. Rupa et al. (2003) has reported that application of farm yard manure in soil elevates the Zn availability in soil to a greater extent. In submerged conditions, the availability of various Zn fractions decreases. The residual fraction is the most dominant Zn fraction and all the fractions appear to be in dynamic equilibrium.

4.4.11   BORON TRANSFORMATION

Tourmaline, a fluorine bromosilicate mineral is the important B-containing mineral. It is highly resistant to weathering, thus making a small fraction of B available to the plants from mineral source. Most of the B in cultivated fields comes from organic matter. Mineralization of organic matter releases B, a portion of it is leached while other portion is held on clay surfaces. Irrigation water is another source of B for crops. B added to soil remains mostly soluble in acidic conditions and is only subjected to leaching in light-textured soil. B adsorption increases with pH, making it unavailable in alkaline soils. Application of organic and inorganic fertilization in soil poses a differential impact on B bioavailability (Diana & Beni, 2006).

4.4.12 MOLYBDENUM TRANSFORMATION

Mo is present largely in the crystal lattice of primary and secondary minerals. Organic matter is a promising source of Mo. It is present in the soil as MoO_4^{2-}. It is either adsorbed on clay complex or present in the soil solution. Unlike other micronutrients, its availability increases with pH (Quaggio et al., 2004). Oxides of Fe and Al increase the adsorption of Mo. Increasing the application of P fertilizers increases the B availability, but S reduces its bioavailability. This is because MoO_4^{2-} and SO_4^{2-} ions are similar in size and charge and they compete at root surface for entry into the plant. An increase in soil moisture enhances Mo availability due to decrease in ferric ion which is responsible for Mo adsorption. Higher concentration of Mo in soil has deleterious effects on the soil fauna (Van Gestel et al., 2011).

4.4.13 NICKEL TRANSFORMATION

Nickel comprises approximately 3% of the earth's crust composition and is the twenty-fourth most abundant element (Lopez & Magnitskiy, 2011). Total Ni concentration commonly ranges from 5 to 500 mg kg^{-1}, with an average of 50 mg kg^{-1} in soils. Ni concentration in dried biosolids is comparatively more than soil. Nickel has five valences: 0, +1, +2, +3, and +4. Among these, only Ni^{2+} is considered an available form for plants. At high pH, Ni^{2+} readily oxidizes and thus becomes unavailable for plant uptake. Additionally, excessive use of Zn and Cu may induce Ni deficiency in soil because these three elements share a common uptake system. Over-liming, which raises pH tremendously, also results in soil to be deficient in plant-available Ni. Thus, in soils that have high pH, either naturally or artificially, Ni fertilization may be needed to ensure good crop quality and yield.

4.5 MECHANISM OF NUTRIENT ABSORPTION BY PLANTS

Plants absorb nutrients from the soil solution via the root network. The absorption of a particular nutrient by plants depends mostly on the availability of that particular element in the soil solution and not decided by the plant itself. The mechanism of nutrient uptake by plants is an important aspect of plant–soil interactions (Morgan & Conolly, 2013). The nutrients are absorbed by the plants in two ways (Robertson, 1951) as described below.

4.5.1 ACTIVE ABSORPTION

Absorption of nutrients from soil solution containing low concentration of nutrients compared to plant cell sap, by expending energy.

4.5.2 PASSIVE ABSORPTION

Nutrients enter the plants along with transpiration stream without the use of energy.

4.5.3 TRANSLOCATION

Translocation of nutrients is a process that involves long-distance transport of plant nutrients from the root zone to the various aerial plant parts via the vascular machinery (Atkins & Smith, 2007). The nutrients are absorbed by root hairs by diffusion. These nutrients enter the cortex and get accumulated in the tissue. From the cortex, the nutrient stream enters the xylem vessels and reaches the leaves by mass flow and transpiration. In case of N uptake, a portion of absorbed NO_3^- is reduced to NH_4^+ and glutamine in roots. These compounds along with remaining portion of NO_3^- pass through the symplast which is the living connection between cells followed by the xylem. From the xylem vessels, they move upward by mass flow in the transpiration stream. Finally, it ends up in leaves where nitrate reduction takes place by the enzyme nitrate reductase (Shaner & Boyer, 1976). The reduced compounds enter phloem vessels and are translocated to growing parts like young leaves, roots, fruits, and others.

4.5.4 ASSIMILATION

Metabolic transformation of inorganic nutrients into organic plant constituents is called assimilation. CO_2 is metabolically transformed into carbohydrates by photosynthesis. Water is assimilated in a few metabolic reactions, but bulk is lost in transpiration. Some nutrients undergo intensive metabolic transformations during their conversion from organic form whereas others do not. A fraction of the absorbed nutrients may be stored in vacuoles without being assimilated. Nutrient assimilation also reflects the use efficiency for various nutrients like N, P, K, and so on. For example, N use efficiency of a particular crop is an important aspect which checks the leaching losses of N into the ecosystem (Masclaux-Daubresse et al., 2010).

4.6   ROLE OF DIFFERENT NUTRIENTS IN PLANTS

4.6.1   NITROGEN (N)

N is a vital plant nutrient and is the most frequently deficient of all nutrients. Plants normally contain 1–50% N by weight. Plants take up N as nitrate (NO_3^-) and ammonium (NH_4^+) ions. Before NO_3^- can be used in the plant, it must be reduced to NH_4^+ or NH_3. Nitrate reduction process differs among the various plant species that employs two enzyme-catalyzed reactions occurring in roots and/or leaves. Both reactions occur in series so that nitrite (NO_2^-) does not accumulate.

$$NO_3^- \xrightarrow{\text{Nitrate reductase}} NO_2^- \text{ (Step 1)}$$

$$NO_2^- \xrightarrow{\text{Nitrite reductase}} NH_3 \text{ (Step 2)}$$

The NH_3 produced in these reactions is used to form numerous amino acids that are subsequently incorporated into proteins and nucleic acids. Nitrate reductase is a water-soluble molybdoflavoprotein. The nitrate reductase enzyme complex exists as a separate subunit which is bound by a molybdenum-binding protein on molecular weight 10,000–20,000 (Fig. 4.3). The mechanism by which the molybdate oxyanion (MoO_4^{2-}) is bound to the enzyme structure is still unknown, and it is the only Mo-enzyme found in higher plants (Goodwin & Mercer, 1998). The enzyme is localized in the cytoplasm or chloroplast and its formation is induced by the plant hormones. The concentration of nitrate reductase varies within the plant species and it depends highly on the organic-N:NO_3^- ratio in xylem exudates. On the other hand, the enzyme nitrite reductase is an intermediate enzyme which involves the conversion of nitrite into NH_3 involving the formation of $-NOH$ and $-NH_2OH$ moieties in between (Fig. 4.4). Nitrite reductase is found mainly in the chloroplast tissue localized in the exterior membrane of the thylakoid membrane.

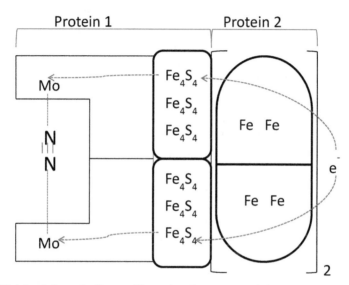

FIGURE 4.3   Schematic diagram illustrating the structure of nitrogenase.

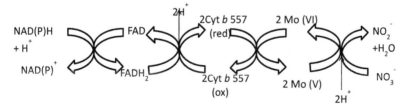

FIGURE 4.4   Schematic diagrams representing the mechanism of nitrate and nitrite reductase action in higher plants.

Proteins provide the structural basis for chloroplasts, mitochondria, and other important cellular organelles that are the sites of various biochemical reactions. Most of the enzymes controlling these metabolic processes are also proteins. These active proteins or enzymes are quite unstable and thus are continually degraded and resynthesized. Apart from its structural importance, N is a vital part of chlorophyll. Chlorophyll's structure can be seen as porphyrin ring system, composed of four pyrrole rings, each containing one N and four C atoms with a single Mg at the center of each porphyrin architecture. Plants receiving required N supply tend to show high photosynthetic activity, vigorous vegetative growth, and a dark green color. The supply of N is related to carbohydrate utilization. At times of N deficiency, carbohydrate deposition occurs in the cells that imparts a thick cellular structure. During the favorable growth period of plants, a required N supply tends to form proteins from the manufactured carbohydrates. Thus, less carbohydrate depositions result in more protoplasm formation making the crops succulent. N fertilization has also been shown to affect the secondary metabolites in plants like phenols, organic acids, essential oils, and so on (Kim & Lee, 2009). Recent reports by Oloyede et al. (2012) and Gasztonyi et al. (2011) portray that inorganic N fertilization has a negative and neutral effect on plant phenols, respectively.

Stunted growth and yellow appearance are some of the popular symptoms of N deficiency. This happens due to loss of protein N from older leaves producing yellowing or chlorosis. Chlorosis appears initially on the lower leaves, the upper leaves remain green, while under severe N deficiency lower leaves turn brown and die. N is a highly mobile element which is visible from its differential physiological symptoms when the young upper leaves remain green and the lower leaves yellow or die. When the roots are unable to absorb sufficient N, protein in the older plant parts is converted to soluble N, translocated to the active meristematic tissues, and used in the synthesis of new proteins.

4.6.2 PHOSPHORUS (P)

Typically, P concentrations in plants vary between 0.1% and 0.4% and are comparatively lower than those of N and K concentrations. Plants absorb P either as $H_3PO_4^-$ or HPO_4^{2-} orthophosphate ions based on the soil pH. At low pH values, the uptake of $H_3PO_4^-$ occurs while HPO_4^{2-} absorption occurs at high pH. During the decomposition of organic matter, plants acquire various organically bound phosphates from soil.

P plays a major role in energy storage and transfer. Adenosine di- and triphosphates (ADP and ATP) act as "energy currency" within plants. When the terminal phosphate molecule from either ADP or ATP is shed off, a large amount of energy (12,000 cal/mol) is put out. This energy obtained from photosynthesis and metabolism of carbohydrates is accumulated in phosphate compounds for subsequent use in plant growth and development.

Increase in root growth is a sign of better P availability in soil. Several other effects on plant growth are attributed to P fertilization. P is related with early maturity of crops, greater straw strength in cereals, and so on. The quality of certain fruits, forage and vegetables, and grain crops improved and disease resistance increased when these crops had satisfactory P nutrition. Cultivation models and quantity of P fertilization significantly influence the expression of phenolic compounds in plants (Oloyede et al., 2012; Pljevljakusic et al., 2014). Application of appropriate amount of P renders frost hardiness to small grains and lowering of root rot diseases. P is a mobile element and so when a deficiency occurs, it is translocated from older tissues to the active meristematic regions. Purple discoloration of the leaves or leaf edges symbolizes P deficiency. Nell et al. (2009) proposed that optimum use of P fertilization and utilization of arbuscular mycorrhizae support good growth of garden sage plant along with the best expression of its metabolite (rosmarinic acid).

4.6.3 POTASSIUM (K)

Plants absorb K as K^+ ion directly from the soil solution by the roots. The range of K^+ concentration in plant tissue varies from 1% to 4% on dry weight basis. K performs functions that are particularly related to the ionic strength of solutions within plant cells.

K plays a pivotal role in enzyme activation processes and thus over 80 plant enzymes require K^+ for their activity. Enzyme activation is regarded as the single most important function of K. Most of these enzymes are found in the meristematic tissues that are undergoing growth process (both above and below ground level), where cell division takes place rapidly and resultantly, primary tissues are formed. For example, the enzyme starch synthetase is involved in the conversion of soluble sugars into starch. Another notable enzyme, nitrogenase, plays a significant role in reducing atmospheric N_2 to NH_3 in the cells of *Rhizobium* sp. Other major roles of K include osmotic regulation of plant cells by which it provides much of the osmotic "pull" that draws water into the plant roots. The mechanism of stomatal opening and closure is facilitated by the K influx in guard cells of leaf. Thus, K regulates transpiration losses of moisture in plants. K is also necessary for the production of ATP. In this way, K aids indirectly in the process of translocation of assimilates. Translocation of sugars requires energy in the form of ATP which requires K for its synthesis. K helps in N uptake and thus in protein synthesis. An incidence of K stress in plants gives rise to crop damage by bacterial and fungal diseases along with pathogenic attacks like insect and mite infestation. Higher application rate of K hampers the production of important phenolic compounds in plants (Nguyen et al., 2010). On the contrary, report by Liu et al. (2011) reveals that K deficiency in soil leads to lowered production of important flavonoids in *Chrysanthemum moriflorum* plants.

4.6.4   CALCIUM (Ca)

Ca is absorbed by plants as Ca^{2+}, and its concentration ranges from 0.2% to 1.0%. Ca has an important role in the structure and permeability of cell membranes. Deficiency of Ca^{2+} promotes disruption of membrane structures, resulting in its loss to retain the important cellular diffusible compounds. The presence of Ca^{2+} also regulates the cation uptake. Ca is essential for cell elongation and cell division. Ca is immobile in nature and hence its deficiency occurs mainly in the storage tissues. Blossom end in tomato and bitter pit in apples are some examples of Ca deficiency. Ca concentration is important for the production of secondary metabolites like phenols, flavonoids, and essential oils in oil palms (Bivi et al., 2014).

4.6.5   MAGNESIUM (Mg)

Mg plays a prominent role in plant physiology. It is absorbed as Mg^{2+}, and its concentration varies between 0.1% and 0.4%. Mg is the primary constituent of the chlorophyll molecule. Apart from this, Mg also serves as a structural component in ribosomes, stabilizing them in the configuration necessary for protein synthesis. Mg^{2+} deficiency causes a significant decline in the proportion of protein N coupled with an increase in nonprotein N in plants. Mg is associated with transfer reactions involving phosphate reactive groups. Mg is required for maximal activity of almost every phosphorylating enzyme in carbohydrate metabolism. Most reactions involving phosphate transfer from ATP require Mg^{2+}.

Mg^{2+} is a mobile element and thus gets easily translocated in the plants. Its deficiency symptoms often appear first on the lower leaves. In many species, shortage of Mg^{2+} results in interveinal chlorosis of the leaves. Sawale et al. (2014) reported the importance of Mg in the formation of phenolic compounds in plants.

4.6.6   SULFUR (S)

S is absorbed by plant roots almost exclusively as the SO_4^{2-} ion. Elemental dusting of fruit plants leads to its absorption whereby it finds its way in small amounts to the integral plant system relatively soon after application. The mechanism of S penetration into the plant is still unknown. Typical S concentration in plants ranges between 0.1% and 0.4%. The S concentration varies among the different crop species. For example, S is present in equal or lesser amounts than P in plants like wheat, corn, beans, and potatoes but in larger amounts in alfalfa, cabbage, and turnips.

S has many important functions in plant growth and metabolism. S supplementation in soil has been reported to enhance the secondary metabolites like glutathione in plants (Fatma et al., 2014). It is required for synthesis of the S-containing amino acids cystine, cysteine, and methionine, which are essential components of protein. Approximately, 90% of the S in plants is locked as amino acids. One of the main functions of S in proteins is the formation of disulfide bonds between polypeptide chains. The disulfide bridge of the amino acids causes folding of the proteins by which it attains a stable structure. S is needed for the synthesis of other metabolites, including coenzyme A, biotin, thiamin, and glutathione. Coenzyme A is of paramount importance in the process of oxidation and synthesis of fatty acids, the synthesis of amino acids, and the oxidation of certain intermediates of the tricarboxylic acid or citric acid cycle. S is also a major component of other S-containing substances, like S-adenosylmethionine, formylmethionine, lipoic acid, and sulfolipid. S is required for the synthesis of chlorophyll. S is a vital part of ferredoxins, a type of nonheme Fe–S protein, occurring in the chloroplasts. Oxidoreduction processes like nitrite reduction, sulfate reduction, the assimilation of N_2 by root nodule bacteria, and free living N-fixing bacteria require ferredoxins. S occurs in volatile compounds and enhances oil production in certain crops. The deficiency symptoms are similar to that of N but the chlorosis is uniform throughout the plant body.

4.6.7 BORON (B)

The B concentration varies in monocots and dicots and the usual concentration ranges from 6 to 18 ppm and from 20 to 60 ppm, respectively. Most of the B is absorbed by plants as undissociated boric acid (H_3BO_3). However, some other forms such as $B_4O_7^{2-}$, $H_2BO_3^-$, HBO_3^{2-}, and BO_3^{3-}, may be present, which generally do not contribute significantly to the plant requirements.

B plays an essential role in the growth and development of new cells in the plant meristem and expression of several flavonoids in plants (Blevins & Lukaszewski, 1998). It is an important element that aids in proper pollination and fruit setting. B helps in the translocation of starch, N, and P. It catalyzes the synthesis of amino acids. Other important functions include nodule formation in legumes, carbohydrate metabolism in plants, and so on. B deficiency is the most widespread micronutrient deficiency. Primary symptom of B deficiency showcases cessation of terminal bud growth that is followed by the death of the young leaves. The deficiency symptoms will appear in the form of thickened, wilted, or curled leaves; a thickened, cracked, or water-soaked condition of petioles and stems; discoloration, cracking, or rotting of fruit, tubers or roots.

4.6.8 IRON (Fe)

Fe is absorbed as Fe^{2+}, Fe^{3+}, and as organically complexed Fe. The sufficiency range of Fe in plant is normally between 50 and 250 ppm. The Fe^{2+} is mostly used by plants for metabolic activities as this form is more mobile and can be easily incorporated into the biomolecular structures.

Fe is an important part of oxidation–reduction reactions in both soil and plants. Due to its chemical properties, Fe is capable of existing in more than one oxidation state and thus it is an integral part of the electron transport system of plant physiological machinery. Fe is needed in many enzymatic transformations (Yeritsyan & Economakis, 2002). Several of these enzymes are involved in chlorophyll synthesis. Thus, a deficiency of Fe leads to chlorophyll malformation or chlorosis. Fe is a structural component of porphyrin molecules, like cytochromes, hemes, hematin, ferrichrome, and leghemoglobin. As much as 75% of the total Fe is associated with the chloroplasts, and upto 90% of the Fe inleaves occurs with lipoproteins of chloroplast and mitochondria membranes. Fe is also a vital part of the enzyme nitrogenase, which is important for N_2 fixation by N-fixing bacteria.

Fe deficiency is most common in crops growing on calcareous or alkaline soils. Fe deficiency first appears in the young leaves of plants. It does not appear to be translocated from older tissues to the tip meristem, and as a result, growth ceases. The young leaves develop an interveinal chlorosis, which progresses rapidly over the entire leaf making it turn entirely white.

4.6.9   MANGANESE (Mn)

Mn concentration in plants typically ranges from 20 to 500 ppm. Mn is absorbed by plants as Mn^{2+}, as well as in molecular combinations with certain natural and synthetic complexing agents.

Mn is intricately involved in the plant photosynthesis and production of several colored metabolites or pigments in plants (Millaleo et al., 2013). It also takes part in oxidation–reduction processes and in decarboxylation and hydrolysis reactions. Mn has a potential to substitute Mg^{2+} in many of the phosphorylating and group transfer reactions. Mn is needed for maximal activity of many enzyme reactions in the citric acid cycle. Mn influences auxin levels in plants, and it seems that high concentration of Mn favors the breakdown of indoleacetic acid.

Like Fe, Mn is a relatively immobile element, and deficiency symptoms usually show up first in the younger leaves. In broad-leaved plants, the symptoms appear as an interveinal chlorosis. Wheat plants low in Mn are susceptible to root rot diseases.

4.6.10   COPPER (Cu)

Cu is absorbed by plants as Cu^{2+}, and may be absorbed as a component of either natural or synthetic organic complexes. Its normal concentration in plant ranges from 5 to 20 ppm. Cu in its reduced form readily binds and reduces O_2. In the oxidized form, Cu is readily reduced, and protein-complexed Cu has a high redox potential. Due to these properties, Cu is utilized in enzymes that create complex polymers such as lignin and melanin. Cu is unique in its involvement in enzymes, and it cannot be replaced by any other metal ion. Various cellular catalytic processes leading to metabolite formation are under direct influence of Cu (Sukalovic et al., 2010).

Deficiencies of Cu have been reported in numerous plants like alfalfa, wheat, barley, oats, beet, lettuce, onions, and others. Symptoms of Cu vary with the crop. Cu-deficient grain plants lose color in the younger leaves, with eventual break and dead tips. In many vegetables, the leaves lack turgor. They develop a bluish green cast, become chlorotic and curl, and flower production fails to take place. Stem melanosis occurs in certain wheat varieties when Cu is deficient.

4.6.11   ZINC (Zn)

The normal Zn concentration in plant tissue ranges from 25 to 150 ppm. Plant roots absorb Zn as Zn^{2+} and as a component of synthetic and natural organic complexes. Soluble Zn salts and Zn complexes can also enter the plant system directly through leaves. Zn is involved in many enzymatic activities, but it is not known whether it acts as a functional, structural, or regulatory cofactor. Zn is important in the synthesis of tryptophane, a component of some proteins and a compound needed for the production of growth hormones (auxins) like indole acetic acid (Mandal et al., 1988).

Zn deficiency can be distinguished by visual symptoms that appear mostly in the leaves, with some unusual occurrences in fruits, branches, or overall plant body. Some of the very common symptoms for Zn deficiency are occurrence of yellow or white areas in the interveinal portions of leaves; production of small, narrow, and thickened leaves; death and early loss of leaves; shortening of stem or stalk internodes resulting in a bushy appearance of the leaves.

4.6.12   MOLYBDENUM (Mo)

Mo is a nonmetal anion absorbed as molybdate (MoO_4^{2-}). This is a weak acid and can form complex polyanions like phosphomolybdate. Normally, Mo content in plants is less than 1 ppm. Mo concentration is low in plants because of the extremely small amounts of MoO_4^{2-} in the soil solution. Mo is an essential component of the enzyme NO_3^- reductase, which catalyzes the conversion of NO_3^- to NO_2^-. Most of the Mo in plants is concentrated in this enzyme, which primarily occurs in leaf chloroplasts. It is also a structural component of the enzyme nitrogenase that is involved in N_2 fixation by root nodule bacteria (Fu et al., 2005). Mo concentration in the nodules of leguminous crops is comparatively higher than in the foliage. Mo is also reported to have an essential role in Fe absorption and translocation in plants. Overall chlorosis is the general symptom of Mo deficiency in plants.

4.6.13   CHLORINE (Cl)

Cl is absorbed by plants as Cl^- ion through both roots and aerial parts. Its normal concentration in plants is about 0.2–20%. Cl^- has not been reported from any plant metabolite (Zhang et al., 2012). The essential role of Cl^- seems to lie in its biochemical inertness. Due to this property, Cl is able to perform osmotic and cation neutralization roles, which may have important biochemical or biophysical consequences. Cl can be easily transported in plant tissues and so it is used as a counter ion during rapid K fluxes, thus contributing to turgor of leaves and other plant parts. Cl plays a definite role in the evolution of O_2 in photosystem II in photosynthesis. Uptake of both NO_3^- and SO_4^{2-} can be reduced by the competitive effects of Cl^-. Many diseases are suppressed by Cl^- fertilizers.

4.6.14   NICKEL (Ni)

Ni is one of the most unique micronutrients because its function in plant growth and development was described in detail before Ni was added to the list of essential elements. The Ni content of crop plants normally ranges from 0.1 to 1.0 ppm dry weight. It is readily taken up by most plants as Ni^{2+}. High levels of Ni may induce Zn or Fe deficiency because of cation competition. Once Ni is absorbed by the root, its movement to the above-ground parts of plants is closely linked to the formation of organic complexes. Nickel is a key component of selected enzymes involved in N metabolism and biological N fixation. Ni is the metal component of urease, the enzyme responsible for degradation of urea and liberation of NH_3. Ni is also associated with the formation of several ureids in plants necessary for protein production (Bai et al., 2006). Apparently, Ni is essential for plants supplied with urea and for those in which ureids are important in N metabolism. Ni is also involved in symbiotic nitrogen fixation through its role as an active center of hydrogenase. The enzyme hydrogenase is responsible for oxidizing the hydrogen produced during symbiotic nitrogen fixation resulting in the production of ATP and, therefore, this enzyme increases the efficiency of symbiotic process, and decreases the inhibitory activity of hydrogen in the bacteroids.

Ni deficiency symptoms include interveinal chlorosis and necrotic spots on young leaves. In general, urea accumulation in the tip of the leaves leading to necrosis of both monocotyledonous and dicotyledonous plants is a visible symptom of Ni deficiency.

4.7 PLANT METABOLITES AND ECOSYSTEM

Plant metabolism greatly relies on the ecosystem functioning and the underlying factors. In spite of some inherent challenges, metabolite profiling is becoming increasingly popular under field conditions. Metabolite studies are used successfully to address topics like species interactions, connections between growth and chemical stoichiometry, and the plant's stress response (Fester, 2015). Ecometabolomics, as this approach has been named, has most often been applied to specific organisms sampled from a given environment. Metabolite levels may be influenced by numerous exogenous or endogenous factors different from the one under study or they may be indifferent to the factor under study due to homeostasis. The very short timescale of metabolic reactions may work in favor of short-term, random factors and to the disadvantage of experimental treatments which usually are effective on a longer timescale.

4.7.1   EXOGENOUS FACTORS

Levels of individual metabolites may be influenced by a large number of external factors like soil parameters or weather conditions. Even without this external variability, experiments using *Arabidopsis thaliana* under phytochamber conditions have demonstrated a relatively high variability in metabolite levels of about 40% when comparing individual plantlets. In field conditions too, the degree of variability is significant for most metabolites. Some metabolites, however, show higher variability while some are less prone to variation with respect to environmental differences. Particularly high variability was observed for some secondary metabolites known to respond to biotic or abiotic stress.

4.7.2   ENDOGENOUS FACTORS

Endogenous factors may exert a significant impact on metabolite profiles, possibly confounding the effects of experimental treatments. A dominant impact of the plant genotype has been documented extensively, although it has not always been encountered. The metabolite profiles in plants are significantly influenced by the plant phenological stages. Moreover, important variations in metabolite profiles also appear among different plant tissues.

4.7.3   HOMEOSTASIS

Homeostasis is the mechanism by which an organism tends to be in harmony and synchronization with its environment. Organisms strive to minimize the impact of external conditions on internal processes and to keep internal conditions (pH, ionic composition, metabolite levels) as stable as possible. This property is mostly visible in trees, where buffering capacities of the plant organism can be particularly strong. However, herbaceous plants may also portray signs of homeostasis. The impact of soil nutrients on metabolite profiles of young plants is comparatively more than in established plants. However in agricultural set ups, the established plants in crop fields get influenced by the environment and the homeostasis exerts a more dominant impact on metabolite profiles. It may be an interesting question, however, whether a specific treatment is able to break plant homeostasis and to significantly affect metabolite levels or not.

KEYWORDS

- **terrestrial plants**
- **life cycle**
- **macronutrients**
- **bioavailability**
- **nitrogen**

REFERENCES

Adesemoye, A. O; Torbert, H. A.; Kloepper, J. W. Plant Growth Promoting Rhizobacteria Allow Reduced Application Rates of Chemical Fertilizers. *Microb. Ecol.* **2009,** *58,* 921–929.

Argen, G. I.; Wetterstedt, J. A. M.; Billberger, M. F. K. Nutrient Limitation on Terrestrial Plant Growth-Modelling the Interaction between Nitrogen and Phosphorus. *New Phytol.* **2012,** *194,* 953–960.

Atkins, C. A.; Smith, P. M. C. Translocation in Legumes: Assimilates. Nutrients and Signaling Molecules. *Plant Physiol.* **2007,** *144,* 550–561.

Bai, C.; Reilly, C. C.; Wood, B. W. Nickel Deficiency Disrupts Metabolism of Ureids, Amino Acids and Organic Acids of Young Pecan Foliage. *Plant Physiol.* **2006,** *140,* 433–443.

Balint, R.; Said-Pullicino, D.; Ajmone-Marsan, F. Copper Dynamics under Alternating Redox Conditions is Influenced by Soil Properties and Contamination Source. *J. Cont. Hydrol.* **2015,** *173,* 83–91.

Bergholm, J.; Olsson, B. A.; Vegerfors, B.; Persson, T. Nitrogen Fluxes after Clear Cutting: Ground Vegetation Uptake and Stump/Root Immobilization Reduce N Leaching after Experimental Liming, Acidification and N Fertilization. *For. Ecol. Manage.* **2015,** *342,* 64–75.

Bivi, M. R.; Farhana, M. D. S. N.; Khairulmazmi, A.; Idris, A. S.; Susilawati, K.; Sariah, M. Assessment of Plant Secondary Metabolites in Soil Palm Seedlings after Being Treated with Calcium, Copper Ions and Salicylic Acid. *Arch. Phytopathol. Plant Prot.* **2014,** *47,* 1120–1135.

Blevins, D. G.; Lukaszewski, K. M. Boron in Plant Structure and Function. *Annu. Rev. Plant Physiol. Plant Mol. Biol.* **1998,** *49,* 481–500.

Boligar, V. C.; Fageria, N. K; He, Z. L. Nutrient Use Efficiency in Plants. *Commun. Soil Sci. Plant Anal.* **2001**, *32*, 921–929.

Brown, P. H.; Welch, R. M.; Cary, E. E. Nickel: A Micronutrient Essential for Higher Plants. *Plant Physiol.* **1987**, *85*, 801–803.

Chapman, N.; Miller, A. J.; Lindsay, K.; Whalley, W. R. Roots, Water, and Nutrient Acquisition: Let's Get Physical. *Trend Plant Sci.* **2012**, *17*, 1360–1385.

Comerford, N. B. Soil Factors Affecting Nutrient Bioavailability. In *Nutrient Acquisition by Plants: An Ecological Perspective*; BasiriRad, H., Ed.; 2005; pp 1–6.

Da Silva Mastos, E.; de Sa Mendosa, E.; Cardoso, I. M.; de Lima, P. C.; Freese, D. Decomposition and Nutrient Release of Leguminous Plants in Coffee Agroforestry System. *R. Bras. Ci. Solo* **2011**, *35*, 141–149.

Diana, G.; Beni, C. Effect of Organic and Mineral Fertilization on Soil Boron Fractions. *Agric. Med.* **2006**, *136*, 70–78.

Epstein, E. The Anolomaly of Silicon in Plant Biology. *Proc. Natl. Acad. Sci.* **1994**, *91*, 11–17.

Estavillo, J. M.; Rodriguez, M.; Gonzalez-Murua, C. Nitrogen Losses by Denitrification and Leaching in Grassland. *Fert. Res.* **1996**, *43*, 197–201.

Fatma, M.; Asgher, M.; Masood, A.; Khan, N. A. Excess Sulfur Supplementation Improves Photosynthesis and Growth in Mustard under Salt Stress through Increased Production of Glutathione. *Environ. Exp. Bot.* **2014**, *107*, 53–63.

Fester, T. Plant Metabolite Profiles and the Buffering Capacities of Ecosystems. Phytochemical **2015**. http://dx.doi.org/10.1016/j.phytochem.2014.12.015.

Fu, G.; Xu, X.; Lu, X.; Wan, H. Mechanisms of Methane Activation and Transformation on Molybdenum Oxide Based Catalysts. *J. Am. Chem. Soc.* **2005**, *127*, 3986–3996.

Gasztonyi, M. N.; Farkas, R. T.; Berki, M.; Petroczi, I. M.; Daood, H. G. Content of Phenols in Wheat as Affected by Varietal and Agricultural Factors. *J. Food Comp. Anal.* **2011**, *24*, 785–789.

Goodwin, T. W.; Mercer, E. I. Nitrogen Fixation, Amino Acid Biosynthesis and Proteins. In *Introductory Plant Biochemistry*, CBS Publishers: New Delhi, 1998; pp 328–361.

Harborne, J. B. General Procedures and Measurements of Total Phenolics. In *Methods in Plant Biochemistry: Plant Phenolics*; Harborne, J. B., Ed.; Academic Press: London, 1989; pp 1–28.

Heidak, M. O.; Glasmacher, U. A.; Scholer, H. F. A Comparison of Micronutrients (Mn, Zn, Cu, Mo, Ni, Na) within Rocks, Soils and Leaves from Fallow Agricultural Lands and Natural Laurel Forest Areas (Tenerife, Canary Islands, Spain). *J. Geochem. Exp.* **2014**, *136*, 55–64.

Helal, F. A. A.; Nagumo, F.; Zewainy, R. M. Influence of Phsosphocompost Application on Phosphorus Availability and Uptake by Maize Grown in Red Soil of Ishigaki Island, Japan. *Agric. Sci.* **2013**, *4*, 102–109.

Kim, D. K.; Lee, K. D. Effects of Nitrogen Application on Growth and Bioactive Compounds of *Chrysanthemum indicum* L. (Gamgug). *Korean J. Med. Crop Sci.* **2009**, *17*, 363–368.

Koorem, K.; Gazol, A.; Opik, M.; Saks, U.; Uibopuu, A.; Sober, V.; Zobel, M. Soil Nutrient Content Influences the Abundance of Soil Microbes but not Plant Biomass at the Small Scale. *PLoS ONE* **2014**. http://dx.doi.org/10.1371/journal.pone.0091998.

Lattenzio, V.; Lattenzio, V. M. T.; Cardinalli, A. Role of Phenolics in the Resistance Mechanisms of Plants against Fungal Pathogens and Insects. In *Phytochemistry: Advances in Research*; Imperato, F., Ed.; Rresearch Signpost: Trivandrum, 2006, pp 23–67.

Lemoine, R.; La Cmera, S.; Atanassova, R.; Dedaldechamp, F.; Allario, T.; Pourtau, N.; Bonnemain, J.-L.; Laloi, M.; Coutos-Thevenot, P.; Maurousset, L.; Faucher, M.; Girrousse, C.; Lemonnier, P.; Parrilla, J.; Durand, M. Source-to-Sink Transport of Sugar and Regulation by Environmental Factors. *Front. Plant Sci.* **2013**. http://dx.doi.org/10.3389/fpls.2013.00272.

Liu, W.; Zhu, D.-W.; Liu, D.-H.; Zhou, W.-B.; Yang, T.-W.; Geng, M.-J. Influence of Potassium Deficiency on Flower Yield and Flavonoid Metabolism in Leaves of *Chrysanthemum morifolium* Ramat. *J. Plant Nutr.* **2011**, *34*, 1905–1918.

Lope-Bucio, J.; Cruz-Ramirez, A.; Herrera-Estrella, L. The Role of Nutrient Availability in Regulating Root Architecture. *Curr. Opin. Plant Biol.* **2003**, *6*, 280–287.

Lopez, M. A.; Magnitskiy, S. Nickel: The Last of the Essential Micronutrient. *Agron. Colomb.* **2011**, *29*, 49–56.

Mandal, B.; Hazra, G. C.; Pal, A. K. Transformation of Zinc in Soils under Submerged Condition and Its Relation with Zinc Nutrition Of Rice. *Plant Soil* **1988**, *106*, 121–126.

Mandal, L. N.; Mitra, R. R. Transformation of Iron and Manganese in Rice Soils under Different Regimes and Organic Matter Applications. *Plant Soil.* **1982**, *69*, 45–46.

Manjunthaiah, B. M. H. Effects of Phosphorus Enriched Organic Manure on Yield and Uptake of Phosphorus in Maize in Vertisol. *Karnatake J. Agric. Sci.* **2003**, *16*, 300–303.

Masclaux-Daubresse, C.; Daniel-Vadele, F.; Dechorgnat, J.; Chardon, F.; Gaufichon, L.; Suzuki, A. Nitrogen Uptake, Assimilation and Remobilization in Plants: Challenges for Sustainable and Productive Agriculture. *Ann. Bot.* **2010**, *105*, 1141–1157.

Meena, M. B.; Biswas, D. R. Phosphorus and Potassium Transformations in Soil Amended with Enriched Compost and Chemical Fertilizers in a Wheat–Soybean Cropping System. *Comm. Soil Sci. Plant Anal.* **2014**, *45*, 624–652.

Millaleo, R.; Reyes-Diaz, M.; Alberdi, M.; Ivanov, A. G.; Krol, M.; Huner, N. P. Excess Manganese Differentially Inhibits Photosystem I Versus II in *Arabidopsis thaliana*. *J. Exp. Bot.* **2013**, *64*, 343–354.

Ming, J.; Xianguo, L; Hongquing, W.; Yuanchun, Z.; Haitao, W. Transfer and Transformation of Soil Iron and Implications for Hydrogeomorphological Changes in Naoli River Catchment, Sanjiang Plain, Northeast China. *Chin. Geogr.. Sci.* **2011**, *21*, 149–158.

Morgan, J. B.; Conolly, E. L. Plant–Soil Interactions: Nutrient Uptake. *Nat. Educ. Know* **2013**, *4*, 2.

Muhammad, S.; Sanden, B. L.; Lampinen, B. D.; Saa, S.; Siddiqui, M. I.; Smart, D. R.; Olivos, A.; Shackel, K. A.; DeJomg, T.; Brown, P. H. Seasonal Changes in Nutrient Content and Concentrations in a Mature Deciduous Tree Species: Studies in Almond (*Prunus dulcis* (Mill.) D. A. Webb). *Eur. J. Agron.* **2015**, *65*, 52–68.

Nell, M.; Votsch, M.; Vierheilig, H.; Steinkellner, S.; Zitterl-Eglseer, K.; Franz, C.; Novak, J. Effect of Phosphorus Uptake on Growth and Secondary Metabolites of Garden Sage (*Salvia officinalis* L.) *J. Sci. Food Agric.* **2009**, *89*, 1090–1096.

Nguyen, P. M.; Kwee, E. M.; Niemeyer, E. D. Potassium Rate Alters the Antioxidant Capacity and Phenolic Concentration of Basil (*Ocimum basilicum* L.) Leaves. *Food Chem.* **2010**, *123*, 1235–1241.

Nieder, R.; Benbi, D. K.; Scherer, H. W. Fixation and Defixation of Ammonium in Soils: A Review. *Biol. Fert. Soils* **2011**, *47*, 1–14.

Oloyede, F. M.; Agbaje, G. O.; Obuotor, E. M.; Obisesan, I. O. Nutritional and Antioxidant Profiles of Pumpkin (*Cucurbita pepo* Linn.) Immature and Mature Fruits as Influenced by NPK Fertilizer. *Food Chem.* **2012**, *135*, 460–463.

Omar, A. E. A. Effect of FYM and Phosphorus Fertilization on Yield and its Components of Maize. *Asian J. Crop. Sci.* **2014**. http://dx.doi.org/10.3923/1cjs.2014.

Partey, S. T.; Quashie-Sam, S. J.; Thevathasan, N. V.; Gordon, A. M. Decomposition and Nutrient Release Patterns of the Leaf Biomass of the Wild Sunflower (*Tithonia diversiflora*): A Comparative Study with Four Leguminous Agroforestry Species. *Agrofor. Syst.* **2011**, *81*, 123–134.

Patrick, W. H. H.; Turner, F. T. Effect of Redox Potential on Manganese Transformation in Water Logged Soil. *Nature* **1968**, *220*, 476–478.

Pljevljakusic, D.; Jankovic, T.; Jelacic, S.; Novakovic, M.; Menkovic, N.; Beatovic, D.; Dajic-Stevanovic, Z. Morphological and Chemical Characterization of *Arnica montana* L. under Different Cultivation Models. *Ind. Crop Prod.* **2014**, *52*, 233–244.

Povilaitis, A.; Sileika, A.; Deelstra, J.; Gaigalis, K.; Baigys, G. Nitrogen Losses from Small Agricultural Catchments in Lithuania. *Agric. Ecosyst. Environ.* **2014**, *198*, 54–64.

Quaggio, J. A.; Gallo, P. B.; Owino-Gerroh, C.; Abreu, M. F.; Cantarella, H. Peanut Response to Lime and Molybdenum Application in Low pH Soils. *R. Bras. Ci. Solo* **2004**, *28*, 659–664.

Robertson, G. P.; Groffman, P. M. Nitrogen Transformations. In *Soil Microbiology, Ecology and Biochemistry*; Paul, E. A., Ed.; Academic Press: London, 2015; pp 421–446.

Robertson, R. N. Mechanism of Absorption and Transport of Inorganic Nutrients in Plants. *Annu. Rev. Plant Physiol.* **1951**, *2*, 1–24.

Rosolem, C. A.; Merlin, A.; Bull, J. C. L. Soil Phosphorus Dynamics as Affected by Congo Grass and P Fertilizer. *Sci. Agric.* **2014**, *71*, 309–315.

Rossi, C. Q.; Pereira, M. G.; Loss, A.; Gazolla, P. R.; Perin, A.; dos Anjos, L. H. C. Changes in Soil C and N Distribution Assessed by Natural δ^{13}C and δ^{15}N Abundance in a Chronosequence of Sugarcane Crops Managed with Pre-Harvest Burning in a Cerrado Area of Goias, Brazil. *Agric. Ecosyst. Manage.* **2013**, *170*, 36–44.

Rupa, T. R.; Rao, A. S.; Singh, M. Effects of Farm Yard Manure and Phosphorus on Zinc Transformations and Phyto-Availability in Two Alfisols of India. *Bioresour. Tech.* **2003**, *87*, 279–288.

Said-Pullicino, D.; Cucu, M. A.; Sodano, M.; Birk, J. J.; Glaser, B.; Celi, L. Nitrogen Immobilization in Paddy Soils as Affected by Redox Conditions and Rice Straw Incorporation. *Geoderma* **2014**, 228–229, 44–53.

Salehi, M. H.; Tahamtani, L. Magnesium Uptake and Palygorskite Transformation Abilities of Wheat and Oat. *Pedosphere* **2012**, *22*, 834–841.

Sanyal, S. K.; De Datta, S. K. Chemistry of Phosphorus Transformations in Soil. In *Advances in Soil Science*; Stewart, B. A., Ed.; Springer-Verlag: Berlin, 1991; pp 1–120.

Sato, S.; Nees, E. G.; Solomon, D.; Liang, B.; Lehmann, J. Biogenic Calcium Phosphate Transformation in Soils Over Millennial Time Scales. *J. Soils Sediments* **2009**, *9*, 194–205.

Sawale, J. A.; Panchal, C. V.; Padmane, S.; Poul, B. N.; Patel, J. R. In Vitro Antioxidant and Anti-Inflammatory Activity of *Wrightia tinctoria* Leaves. *World J. Pharm. Pharmaceut. Sci.* **2014**, 964–972.

Scott, J. T.; Lambie, S. M.; Stevenson, B. A.; Schipper, L. A.; Parfitt, R. L.; McGill, A. C. Carbon and Nitrogen Leaching under High and low Phosphate Fertility Pasture with Increasing Nitrogen Inputs. *Agric. Ecosyst. Environ.* **2015,** *202,* 139–147.

Shaner, D. L.; Boyer, J. S. Nitrate Reductase Activity in Maize (*Zea mays* L.) Leaves. *Plant Physiol.* **1976,** *58,* 499–504.

Sharma, K. L.; Bajaj, J. C.; Das, S. K.; Rao, U. M. B.; Ramalingaswami, K. Nutrient Transformation in Soil Due to Addition of Organic Manure. *Fertil. Res.* **1992,** *32,* 303–311.

Silver, W. L. Is Nutrient Availability Related to Plant Nutrient Use in Humid Tropical Forests? *Oecologia* **1994,** *98,* 336–343.

Sukalovic, V. H.-T.; Vuletic, M.; Veljovic-Jovanovic, S.; Vucinic, Z. The Effects of Manganese and Copper In Vitro and In Vivo on Peroxidase Catalytic Cycles. *J. Plant Physiol.* **2010,** *167,* 1550–1557.

Svoboda, N.; Taube, F.; Wienforth, B.; Kluß, C.; Kage, H.; Herrmann, A. Nitrogen Leaching Losses after Biogas Residue Application to Maize. *Soil Tillage Res.* **2013,** *130,* 69–80.

Tate, K. R. The Biological Transformation of P in Soil. *Plant Soil.* **1984,** *76,* 245–256.

Tester, M.; Leigh, R. A. Partitioning of Nutrient Transport Processes in Roots. *J. Exp. Bot.* **2001,** *52,* 445–457.

Tisdale, S. L.; Nelson, W. L.; Beaton, J. D.; Havlin, J. L. *Soil Fertility and Fertilizers,* 5th ed.; Prentice Hall of India: New Delhi, 1995.

Tunesi, S.; Poggi, V.; Gessa, C. Phosphate Adsorption and Precipitation in Calcareous Soils: The Role of Calcium Ions in Solution and Carbonate Minerals. *Nutr. Cycl. Agroecosyst.* **1999,** *53,* 219–227.

Van Gestel, C. A. M.; Borgman, E.; Verweij, R. A.; Ortiz, M. D. The Influence of Soil Properties on the Toxicity of Molybdenum to Three Species of Soil Invertebrates. *Ecotox. Environ. Saf.* **2011,** *74,* 1–9.

Vandre, R.; Clemens, J. Studies on the Relationship between Slurry pH, Volatilization Processes and the Influence of Acidifying Additive. *Nutr. Cycl. Agroecosyst.* **1997,** *47,* 157–165.

Velarde, M.; Felker, P.; Gardiner, D. Influence of Elemental Sulfur, Micronutrients, Phosphorus, Calcium, Magnesium and Potassium on Growth of *Prosopis alba* on high pH Soils in Argentina. *J. Arid Environ.* **2005,** *65,* 525–539.

Vong, P.-C.; Nguyen, C.; Guckert, A. Fertilizer Sulphur Uptake and Transformations in Soil as Affected by Plant Species and Soil Type. *Eur. J. Agron.* **2007,** *27,* 35–43.

Xiang, H. F.; Tang, H. A.; Ying, Q. H. Transformation and Distribution of forms of Zinc in Acid, Neutral and Calcareous Soils of China. *Geoderma* **1995,** *66,* 121–135.

Yeritsyan, N.; Economakis, C. Effect of Nutrient Solution's Iron Concentration on Growth and Essential Oil Content of Oregano Plants Grown in Solution Culture. *Acta Hortic.* **2002,** *576,* 278–283.

Yildirin, E.; Karlidog, H.; Turan, M.; Dursun, A.; Goktepe, F. Growth, Nutrient Uptake, and Yield Promotion of Broccoli by Plant Growth Promoting Rhizobacteria with Manure. *Hortic. Sci.* **2011,** *46,* 932–936.

Zhang, S.; Asadullah, M.; Hocking, R.; Lin, J.–Y.; Li, C.-Z. Transformation of Chlorine in NaCl-Loaded Victorian Brown Coal during the Gasification in Steam. *J. Fuel Chem. Tech.* **2012,** *40,* 1409–1414.

CHAPTER 5

SECONDARY PLANT METABOLITES: MECHANISMS AND ROLES IN INSECT PEST MANAGEMENT

TAMOGHNA SAHA[1]*, NITHYA C.[2], SHYAMBABU S.[1], KIRAN KUMARI[1], S. N. RAY[1], and KALMESH M.[1]

[1]Department of Entomology, Bihar Agricultural University, Sabour, Bhagalpur 813210, Bihar, India

[2]Division of Entomology, Indian Agricultural Research Institute (IARI), New Delhi 110012, India

*Corresponding author, E-mail: tamoghnasaha1984@gmail.com.

CONTENTS

ABSTRACT

Food requirement has forever been a challenge for mankind. Usually, in all natural habitations, plants are encircled by an immense number of biotic and diverse categories of abiotic environmental stress. Virtually, flora and fauna contain a large diversity of bacteria, viruses, fungi, nematodes, mites, insects, mammals, and other herbivorous animals, wholly accountable for heavy reduction in crop productivity. To protect from herbivory and disease-causing organisms, plants have devolved an excess of secondary metabolites which have given management against pest. They might have performed as repellents, antifeedants, phagostimulants, or toxins against the pests. The worldwide inclination to minimize the misuse of toxic materials on crops has led to several studies focusing on alternative pest management technologies for crop fortification. Thus, studies of secondary plant chemicals have been increasing day by day. These secondary metabolites, including terpenes, phenols, and nitrogen (N)- and sulfur (S)-comprising compounds, guard plants against a diversity of herbivores and pathogenic microorganisms with diverse kinds of abiotic pressures. This chapter will give an impression about a few of the methods by which plants defend themselves against herbivory, pathogenic microbes, and various abiotic strains as well as specific plant reaction to pathogen attack.

5.1 INTRODUCTION

In this present world, insects are the most divergent group of animals on the earth, including more than half of all known living organisms. Even as most of insects rely on plants' primary metabolites (e.g., starch, proteins, and lipids), plants have also developed a high range of plant metabolites (e.g., alkaloids, terpenoids, and phenolics) to cope with heavy herbivory. Arthropods and pathogens are constantly challenged with counter mechanisms to detoxify plants' defense systems. Another side, arthropods, and shrubs and herbs have coevolved mutualistic relationships through pollination. In such a complex ecological network, insects have gradually developed highly responsive and specific chemical sensors and biochemical processes to detect allelochemicals—to find host plants and nectar sources, to avoid noxious plants, and to escape from predators, with their intraspecific recognition via pheromonal communications. However, the use of plant-derived products to manage the insect pest is a well-known fact in the rising world, and before the invention of synthetic pesticides, plant-derived products were only pest-managing representatives to farmers all over the world (Ritsuo, 2014; Owen, 2004).

As we know, plants produce enormous and broad series of phytochemicals within their system, but not all these compounds directly participate in the expansion and progress of the plant. These phytochemicals, also recognized as plant metabolites, are disseminated largely surrounded by the plant realm and the plant physiological make-up. Plant derived chemical structure and biosynthetic paths are well known for their complications and these phytochemicals have been widely investigated for their chemical properties for an extended point of time. The current focus of the phytochemicals is the search of their biological properties of new drugs/ medicines, antibiotics, insecticides, herbicides, and behavior-modifying chemicals. Based on literature information, most of these composites have given importance in protection against herbivorous nature, and the phytochemical diversity of insect defenses in tropical and temperate plant families has too been significantly proven (Croteau et al., 2000).

The majorities of these secondary metabolites show indirect roles in growth of the plant and its development and are significant to those plants that show their biosynthesis (Taiz & Zeiger, 2006). Due to toxicity of plant metabolites, they show protective function against biotic factors like defense from attack of disease-causing organisms and arthropods, allelopathy, and so on (Athanasiadou & Kyriazakis, 2004; Khan & Singh Rattan, 2008).

Hence, for these environmental security and fitness matters (Karunamoorthi et al., 2008), currently, new strategies for pest insects control which are less hazardous and ecofriendly to the atmosphere are searched; such as plant products show environmental friendly methods for this aim (Salvadores et al., 2007).

However, more than the last five decades, insect pests have mainly been controlled by chemical pesticides. Therefore, resistance problems are happening against pesticide and adverse result goes to nontarget organisms, including human being and the environment, opposing the acceptance of the use of these synthetic compounds. Injudicious application of synthetic pesticides causes serious problems in environment, including genetic resistance of pest species, residual problem in stored products, risk to handle, environmental contamination, and so on (Rembold, 1994; FAO, 1992). The plant-based insecticides are usually pest-specific and are comparatively safe to nontarget organisms including man and are moreover ecofriendly and safe to the environment.

The plants that showed biological activity against insects owe this feature to the existence of phytochemicals (Garcia et al., 2004). There is a process of acquiring plant metabolites from botanical extracts against insects, which can have deviations; sometimes aqueous extracts can be acquired (Bobadilla et al., 2005), solvents can be applied to achieve different compounds that depend on their polarity (Bobadilla et al., 2002), vital oils can also be acquired (Perez-Pacheco et al., 2004), or dried out plant parts to be used as powders (Silva et al., 2003).

However, secondary plant metabolites attribute numerous properties against insects, like insecticidal action (Cavalcante et al., 2006), considered as that matter or combination of substances that perform biocide action owing to the character of their chemical structure (Celis et al., 2008). Though, the greater part of the plant extracts utilized against pests have an insectistatic result, rather than insecticidal. This refers to the obstruction of the insect's development and behavior (Celis et al., 2008), and it is divided into antifeedent, repellence (Viglianco et al., 2006), growth parameter (Wheeler & Isman, 2001), feeding restriction (Koul, 2004), and oviposition deterrents (Banchio et al., 2003).

However, the objective of the current chapter is to know the effect of plant metabolites and pure allelochemicals and to discuss their implication for insect pest control.

5.2 SECONDARY METABOLITES AS CHEMICAL BARRIERS FOR INSECTS

In addition, insecticidal and other noxious plant allelochemicals have a wide range of secondary plant metabolites to interrupt processes such as insect host-finding behavior and endocrinological systems. However, plants can differentiate between general cutting and feeding by the insects in the existence of elicitors in the saliva of chewing insects. Plants discharge volatile organic compounds (VOCs) by response and those compounds are monoterpenes, sesquiterpenes, and homoterpenes. Such compounds may keep the harmful insects away or evoke beneficial predators that prey on the destructive pests. Some of the examples like wheat seedlings affected by aphids may fabricate VOCs that deter other aphids. Lima beans and apple trees release some chemical compounds that lure predatory mites when wounded by spider mites, and cotton plants manufacture volatiles that evoke predatory wasps when damaged by moth larvae. The undamaged portion of the plant tissues can persuade systemic production of these chemicals, and once they are released, they perform as signals to adjacent plants to start producing similar compounds. The extracts of such chemical compounds have highest metabolic price on the host plant, so most of these composites are not formed in huge quantities until after insects have begun to feed (Modupe & Musa, 2014).

5.3 SECONDARY PLANT METABOLITES AS DEFENSE FOR INSECT PEST MANAGEMENT

Plant metabolites are formed in two main components—primary metabolites and secondary metabolites. The primary metabolites consist of whole plant cells elements that are straightly involved in growth, or reproduction processes; these comprise sugars, proteins, amino acids, and nucleic acids. Secondary metabolites are those plant elements that are indirectly concerned in the growth of the plants but are occupied in the physiological aspect including the plant defense system. A few are indisputably produced for easily appreciated reasons, for example, as toxic materials giving protection against predators, as volatile attractants on the way to the same or other species, or as coloring agents to trap or warn other species, but it is reasonable to presume that all do play some imperative for the well-being of the producer. The plant metabolites in high concentrations might result in a more resistant plant (Modupe & Musa, 2014). Furthermore, some essential classes of secondary metabolites were also used in commercial botanical pest management (Table 5.1). Hence, defense plant metabolites could be separated into constitutive matters, also called prohibitins or phytoanticipins and stimulated metabolites that are produced in response to an infection relating *de novo* enzyme synthesis, known as phytoalexins (Van Etten et al., 1994; Grayer & Harborne, 1994). High energy and carbon-consuming phytoanticipins show fitness cost under natural conditions (Mauricio, 1998), but established as the foremost line of chemical defense that powerful pathogens have to overcome. In contrast, the manufacturing of phytoalexin may require 2 or 3 days, at opening the enzyme system needs to be synthesized (Grayer & Harborne, 1994).

TABLE 5.1   Some Examples of Botanical Pest Management Compound Based on Most Important Classes of Plant Compound.

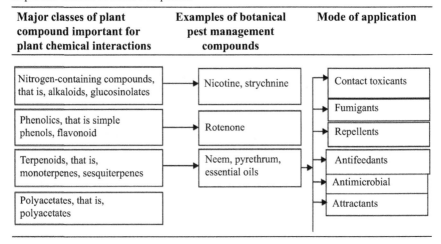

Major classes of plant compound important for plant chemical interactions	Examples of botanical pest management compounds	Mode of application
Nitrogen-containing compounds, that is, alkaloids, glucosinolates	Nicotine, strychnine	Contact toxicants
		Fumigants
Phenolics, that is simple phenols, flavonoid	Rotenone	Repellents
Terpenoids, that is, monoterpenes, sesquiterpenes	Neem, pyrethrum, essential oils	Antifeedants
		Antimicrobial
Polyacetates, that is, polyacetates		Attractants

5.4   MOST IMPORTANT SET OF PLANT METABOLITES

Principal secondary plant metabolites could be separated into four individual groups, namely, terpenoids, phenolics, alkaloids, and sulfur-containing compounds. Some specific examples which perform as insect antifeedants cutoff from terrestrial plants are mentioned.

5.4.1   TERPENOIDS (TERPENES)

These groups are established in the majority of the plants and corresponded to the majority important class of secondary plant metabolites with estimated number of recognized and isolated chemical compounds in the series of about 22,000. The simplest example of terpenoids, the hydrocarbon isoprene (C_5H_8) is a volatile gas released throughout photosynthesis in huge quantities by leaves that may defend cell membranes from injury caused by high temperature or light. These are grouped by the number of isoprene units utilized to construct them (Hassan, 2011). Terpenes are the largest class of secondary plant metabolites and are combined by their familiar biosynthetic origin from acetyl-coA or glycolytic intermediates (Gershenzon & Croteau, 1991). A vast common of the diverse terpenes compositions formed by plants as metabolites that are supposed to be mixed up in defense as toxins and antifeedents to a huge amount of plant feeding arthropods and mammals (Gershenzon & Croteau, 1991). The subclasses of terpenoides are mentioned below.

5.4.1.1   MONOTERPENES (C10)

Monoterpenes (based on a 10-carbon frame) which are most important elements of numerous plant "essential oils" prevent insect feeding. Essential oils frequently perform as insect poisons and give protection from fungal or bacterial attack. Mints (*Mentha* spp.) generate huge amount of the monoterpenoids menthol and menthone which are created and stored up in glandular trichomes on the epidermis (Hassan, 2011). The other terpenoids (hydrocarbons having oxygen) are acyclic monoterpene alcohols (geraniol and linalool), monocyclic alcohols (menthol, 4-carvomenthenol, terpineol, carveol, and borneol), aliphatic aldehydes (citral, citronellal, and perillaldehyde), aromatic phenols (carvacrol, thymol, safrole, and eugenol), bicyclic alcohol (verbenol), monocyclic ketones (menthone, pulegone, and carvone), bicyclic monoterpenic ketones (thujone, verbenone, and fenchone), acids (citronellic acid and cinnamic acid), and esters (linalyl acetate). A few crucial oils may also have oxides (1,8-cineole), sulfur-containing elements, methyl anthranilate, coumarins, and so on. Zingiberene, termerone, curcumene, farnesol, sesquiphellandrene, nerolidol, and others, are paradigms of sesquiterpenes (C15) cutoff from essential oils. The plant phenolics are also a good source of bioactive compounds (Macas et al., 1999).

Majority of the monoterpenes imitative are vital agents of insect toxicity. For example, the pyrethroids (monoterpenes esters) occurring in the leaves and flowers of *Chrysanthemum* species show strong insecticidal responses (neurotoxin) to insects like beetles, wasps, moths, bees, and others, and a well-known ingredient in marketable insecticides because of little diligence in the atmosphere and low mammalian toxicity (Turlings et al., 1995). However, in gymnosperms like Pine and Fir, accumulation of monoterpenes in the ducts resin found in the needles, twigs, and trunks mostly as α-pinene, β-pinene, limonene, and myrecene, all are toxic to various insects like bark beetles, severe pest of conifer species all over the world (Turlings et al., 1995). Many monoterpenes have strong biological activity against pests and have been commercially exploited throughout the earlier decade and some have been marketed by EcoSmart in the USA (Koul et al., 2008).

5.4.1.2   DITERPENES (C20)

Numerous types of diterpenoides (based on a 20-carbon skeleton) are recognized as antifeedants, including the clerodanes and the abietanes. Amongst the diterpenoids, gossypol produced by cotton (*Gossypium hirsutum*) has shown to be a sturdy antifeedant, and possesses antifungal and antiseptic properties (Hassan, 2010). However, neo-clerodane diterpenes show potential group of compounds that affect the feeding behaviour of insect pests. Approximately, 150 neo-clerodanes have been cutoff (Ortego et al., 1995) and among these, eriocephalin and teucvin are quite effective along with ajugarins isolated from *Ajuga remota* (Kubo et al., 1982; Kubo et al., 1983). Compounds resembling ajugarins such as ajugareptansin and ajuga reptanoside-A and -B from *Ajuga reptans* (Camps et al., 1981) and *Ajuga riva* (Camps et al., 1982) are also significant anti-insect allelochemicals.

In addition to neo-clerodane diterpenoids, cutoffs from various species of teucridium, ajuga, and scutellaria (Family Lamiaceae) also inhibit feeding in lepidopteran larvae. In skullcap (*Scutellaria galericulata*), aerial part, jodrellin-T, 14,15-dihydro jodrellin-T, and galericulin are novel structures (Koul, 2004). Jodrellin-B, also reported from *Scutellaria woronowii*, is the most active compound in this series and scutalpin-C from *Scutellaria alpina javalaambrensis* is very lively against *Spodoptera littoralis* (Munoz et al., 1997). Together, clerodane and neo-clerodane group of diterpenoids are well identified for their insecticidal (Camps & Coll, 1993) and antifeedant activity (Cole et al., 1990; Belles et al., 1985). Clerodane diterpenes, 3,13E-clerodane-15-oic acid, 4,13E-clerodane-15-oic acid, 18-oxo-3,13E-clerodane-15-oic acid, and 2-oxo-3,13E-clerodane-15-oic acid from the Nigerian plant *Detarium microcarpum* perform as feeding deterrents (Lajide et al., 1995a), particularly in opposition to workers of *Reticulitermes speratus* (subterranean termite). The extraordinarily hard wood of a Nigerian plant, *Xylopia aethiopica*, also survives attack from termites and other insects destructive to wooden structures; this has led to the isolation of ent-kauranes, (7) kaur-16-en-19-oic acid which has a strong feeding deterrent activity against termite workers of *R. speratus Kolbe* (Lajide et al., 1995b). Numerous natural neo-clerodane diterpenoids cutoff from *Linaria saxatilis* and some semisynthetic imitative was tested against numerous insect species with different feeding adaptations. The antifeedant tests found that the Colorado feeding beetle, *Leptinotarsa decemlineata*, an oliphagous insect was the major responsive insect, followed by the aphid *Myzus persicae*. The polyphagous *S. littoralis* was not dissuaded by these diterpenoids; however, subsequent oral management, a few of these compounds did have post-ingestive antifeedant effects on this insect (Gonzalez et al., 2000).

In contrast, abietic acid is a diterpene observed in pines and leguminous trees. It is there in or along with resins in resin canals of the tree trunk. When these canals are penetrated by feeding insects, the resin outflow may physically obstruct feeding and perform as a chemical deterrent to continued predation (Bradley et al., 1992). Furthermore, gibberellins, a plant hormone group, are also known as diterpenoids, that take part in a series of harmful roles in several plant developmental processes such as seed germination, leaf expansion, setting of flower and fruit (Davies, 1995), dry mass, and biomass production (Gupta et al., 2001).

5.4.1.3   TRITERPENES (C30)

Many of well-recognized insect antifeedants are triterpenoids. Based on a 30-carbon skeleton, these substances frequently appear as glycosides (conjugated with sugars) and are normally highly oxygenated. Particularly well considered in this view are the limonoids from the neem (*Azadirachta indica*) and chinaberry (*Melia azedarach*) trees, demonstrated by azadirachtin, toosendanin, and limonin from Citrus species. Other antifeedant triterpenoids include cardenolides, steroidal saponins, and withanolides (Hassan, 2010). They are parallel in molecular formation to plant and mammal sterols and steroid hormones. Phytoectysones are imitates of insect-molting hormones. When manufactured by plants such as spinach (*Spinacia oleracea*), they interrupt larval growth and increase insect mortality. The fresh odor of lemon and orange skin is the effect of a class of triterpenes called limonoids. Essential oil of citronella separated from lemon grass (*Cymbopogon citratus*) also includes huge limonin levels and has become a remarkable insect repellent in the USA owing to its low toxicity in humans and biodegradable properties. The caterpillars of monarch butterfly feed approximately completely upon milkweed (*Asclepias* spp.) which presents huge quantities of these poisons in the milky latex of their sap. The caterpillars retain these poisons carefully within their bodies, and when they become adult butterflies, they are highly noxious to the majority of predatory birds that consume them. This aids with stopping the birds from aggressive feeding on preharvest crops on the farms. Saponins are glycosylated triterpenes (triterpenes with attached sugar groups), which exist in the cell membranes of numerous plant species. These materials have soap-like properties (cleansing agent) and interrupt the cell membranes of occupying fungal pathogens (Freeman & Beattie, 2008). In addition, numerous steroid alcohols (sterols) are an important part of plant cell membranes, particularly in the plasma membrane as they are control channels and allow the motion of the fatty acid chains. The milkweeds fabricate numerous improved flavors of glucosides (sterols) that defend them against herbivory such as a majority of the insects and even cattle (Lewis & Elvin-Lewis, 1977). One more set of triterpene is limonoid, a bitter material found in citrus fruits that performs against herbivores in members of family Rutaceae and some other families also. For example, azadirachtin, a compound of limonoid from *Azadirachta indica*, performs as an antifeedent to some insects and applies diverse deadly effects (Mordue & Blackwell, 1993).

5.4.1.4 SESQUITERPENES (C15)

Based on 15-carbon skeleton, these substances act as strong antifeedants like drimanes (e.g., drimane polygodial, which occurs in plants of the water pepper, *Polygonum hydropiper*) and sesquiterpenoid lactones. However, sesquiterpenoids are the most important parts of essential oils, which are extremely volatile composites giving out the aroma (essence) of plants that create them. These sets are a vital source of insect antifeedants (Hough-Goldstein, 1990). Many insecticidal and antifeedant sesquiterpenoids are known as major deterrents in insect–plant interactions (Rodriguez, 1985; Rossiter et al., 1986; Ivie & Witzel, 1982). Another tricyclic silphinene, sesquiterpene, namely 11β-acetoxy-5α-(angelyloxy)-silphinen-3-one, and two of its hydrolytic products, 11β-hydroxy-5α-(angeloyloxy)-silphenen-3-one and 11β,5-dihydroxy-silphinen-3-one, reported from *Senecio palmensis* (Asteraceae) are considered as strong antifeedant compounds against the Colorado potato beetle, *Leptinotarsa decemlineata* (Say) (Eigenbrode et al., 1994). Siliphene sesquiterpenes, however, have both antifeedant and poisonous result against insects.

Some other feeding deterrents such as sesquiterepene lactone angelate argophyllin-A and 3-*O*-methyl-niveusin-A have been cutoff from inflorescences of cultivated sunflower. a-Cyperone, a sesquiterpene cutoff from the *Cyperus rotundus* (nutgrass) tubers is insecticidal against diamondback moth *Plutella xylostella* (Thebtaranonth et al., 1995). Drimane group of sesquiterpenes occurs in the marsh pepper *P. hydropiper* (Polygonaceae), besides in the plants of the genera *Warburgia, Cinnamosma, Winterana,* and *Cinnamodendron* (Cannellaceae). Poligodial, warburganal, and muzigadial are among a few of the potential drimane sesquiterpenes having anti-insect and opposing fungal properties (Jansen & Groot, 1991). Inhibition of nourishing in monophagous with polyphagous insects has been attributed to enal and α, β unsaturated aldehyde group(s) in these molecules (Kubo et al., 1978).

Costunolide, the bioactive sesquiterpene lactone, is a 5-membered lactone ring (a cyclic ester) and has powerful nourishing that keeps away many herbivorous insects as well as mammals (Picman, 1986). The root bark of Chinese bittersweet *Celastrus angulatus* is traditionally utilized in China to defend foliages from harmful insects and contains polyol ester celangulin that deters feeding in insects. This compound has a dihydro-agarofuran skeleton with seven hydroxyl functions, five acylated, one benzoylated, and one free (Wakabayashi et al., 1988). Abscisic acid also performs as sesquiterpene, serving primarily regulatory roles in the beginning and then protection of seed and bud dormancy and helps plants reply to water tension by adjusting the membrane properties (Van Steveninck & Van Steveninck, 1983). A few current records also demonstrate that such terpenes cutoff recently from *Rutales* are effective feeding deterrents for stored grain pests, particularly the spirocaracolitones, which are complete antifeedants (Omar et al., 2007).

5.4.1.5   POLYTERPENES (C5)n

Numerous high molecular weight polyterpenes occur in plants. Larger terpenoids include the tetraterpenes and the polyterpenes. The most important tetraterpenes are carotenes family of pigments. One of them is rubber, a polymer holding 1500–15,000 isopentenyl units, in which almost all the C–C double bonds have a *cis*(Z) pattern whilst in gutta rubber has its dual bond in *trans*(E) configuration. Rubber which usually gets established in long vessels called laticifers helps in curing of wounds and provides defense against herbivores (Eisner and Meinwald, 1995; Klein, 1987).

5.5   ALKALOIDS

Alkaloids obviously occur in organic compounds having nitrogen moiety and are observed in many vascular plants and include caffeine, cocaine, morphine, and nicotine. They originate from the polypeptides aspartate, lysine, tyrosine, and tryptophan, and a number of these materials have potent effects on animal physiology. Alkaloids with familiar antifeedant consequence on insects comprise specific indoles and the solanaceous glycol alkaloids. The associates of the Solanaceae family manufacture many vital alkaloid composites. Amongst them nicotine is one of the vital alkaloids, produced from the roots of tobacco plants (*Nicotiana tabacum*) and transferred to foliage where it is stored in vacuoles. This is discharged when arthropods nibble on the leaves and break open the vacuoles. Another alkaloid is caffeine which is observed in foliage such as coffee (*Coffea arabica*), tea (*Camellia sinensis*), and cocoa (*Theobroma cacao*). This is toxic to both insects and fungi. Indeed, immense levels of caffeine manufactured by coffee saplings could still hamper the growth of other seeds in the surrounding area of the growing plant, a phenomenon called allelopathy. Allelopathy permits single plant species to "defend" itself against other plants that may compete for growing space and nutrient resources (Hassan, 2011; Modupe & Musa, 2014).

However, new alkaloidal compounds, (+)-11-methoxy-10-oxoerysotramidine and (+)-10,11-dioxoerysotramidine, recognized in the seeds, seed pods, and blossoms of *Erythrina latissima* are found to have strong antifeedant activities against third-instar *S. littoralis* (Boisduval) larvae.

Erythrinaline alkaloids having antifeedant effect show their effective-ness in postharvest maintenance and crop protection. It was also found that farms with maize cultivated under the *E. latissima* tree were scarcely attacked by the stem borer. Since the tree is a widespread flowering plant, its seeds and flowers can be bound and utilized as strong biopesticides or antifeedants in agricultural postharvest processes (Wanjala et al., 2009). Atropine is a neurotoxin and cardiac stimulant produced by the deadly nightshade, *Atropa belladonna*. Although it is poisonous in huge quan-tities, it has been utilized curatively by humans in little quantities as a student dilator and remedy for some nerve gas toxins.

Other plant metabolites with antifeedant effects comprise cyanogenic glycosides, another secondary metabolite from plants also known to have toxic nitrogenous compounds accountable for the decomposition of hydrogen cyanide (HCN) which stops cellular respiratory action in aerobic organisms. Cyanogenic glycosides which are produced from plants, also manufacture enzymes that convert composites into HCN, including glyco-sides and hydroxynitrile lyases. However, these are stored in different sections or tissues, that is within the foliage. When arthropods feed on these tissues, their enzymes and substrates mix up and generate deadly HCN (Freeman & Beattie, 2008). Some other compounds belonging to the aliphatic ketones and aldehydes of *Commiphora rostrata*, have demon-strated great repellent action against *Sitophilus zeamais* in contrast to the not genuine commercial repellent of insect pests *N,N*-diethyl toluamide [DEET] used in the study (Lwande et al., 1992). Essential oils of *Matricaria recutita* identified to have precocenes as an active constituent also inter-fere with glands of pests that manufacture juvenile hormones resulting in the restraint of insect development while molting (Ndungu et al., 1995). The plant species discharge a diversity of volatiles including various alco-hols, terpenes, and aromatic compounds that supply as defense mecha-nisms against arthropods and pathogens. These elements stop insects or other herbivores from feeding, and produce straight toxic effects, or get mixed up in engaging biocontrol agents in response to feeding damage. They could also be utilized by the plants to draw pollinators, protect plants from disease, or help in interplant communication (Ndungu et al., 1995; Pichersky & Gershenzon, 2002).

However, the use of green pesticides mainly for stored grain pests has come with high recommendations in current moments and the essential oils appear to be the best choices being suggested. Studies have revealed that essential oils are readily biodegradable and less harmful to pollinators and untargeted organisms as compared to synthetic pesticides (Baysal, 2008). Application of foliage products particularly essential oils is an extremely eye-catching method for controlling post-harvest crop diseases. Production of essential oils by plants is supposed to be principally a defensive method against arthropod pests and disease-causing organisms (Oxenham, 2008). Nowadays, essential oils are attaining considerable acceptance for use inside postharvest maintenance, crop protection, and fumigation because of the relative safety, biodegradable status, and their utilization for other multipurpose uses.

5.6 PHENOLS

Plants produce an enormous range of secondary plant metabolites that hold a phenol group, a hydroxyl functional group on an aromatic ring called phenol, also a chemically heterogeneous group. These composites are an important group of secondary metabolites observed within plant tissues that provide as protection mechanism against pathogens, produced through the shikimic acid and malonic acid pathways in plant systems. They cover a wide series of defense-related plant metabolites, for example, lignin, tannins, flavonoids, anthocyanins, and furanocoumarins (Modupe & Musa, 2014). Phenolics might be a vital component of the foliage defense structure against insects and pathogens with root parasitic nematodes (Wuyts et al., 2006). Each and every compound under phenolic group is explained below.

5.6.1 LIGIN

They are well-branched heterogeneous polymer of phenyl-propanoid groups, produced by three diverse alcohols, namely, coumaryl, coniferyl, and sinapyl which are oxidized to free radicals (reactive oxygen species [ROS]) by a ubiquitous plant enzyme—peroxidase, and reacts simultaneously and randomly to form lignin. The three monomeric elements in lignin with their reactive proportions differ amongst species, plant organs, and still layers of a distinct cell wall (Lewis & Yamamoto, 1990). Hundreds of thousands of phenolic monomers of lignin group have primary component of wood. Because it is insoluble, stiff, and almost hard to digest, lignin

gives an excellent physical barrier against pathogen attack. The physical toughness of lignin prevents feeding by herbivorous animals and its
chemical stability makes it comparatively indigestible to herbivores and
insects pathogens (Mader & Amberg-Fisher, 1982). Lignifications limit
the expansion of disease-causing organisms and are a repeated response to
infection or wounding (Gould, 1983).

5.6.2 TANNINS

Tannins are soluble in water flavonoid polymers formed by foliage and
stored in vacuoles. These are included under the second class of plant
phenolic polymers through protective properties. Mainly tannin has
molecular masses between 600 and 3000. They are common toxins that
considerably prevent the development of many herbivores and also serve
as feeding repellents to an enormous range of animals. They are also
noxious to insects as they join to salivary proteins and digestive enzymes
with trypsin and chymotrypsin effect in protein inactivation (Hassan,
2011). The insect arthropods that consume huge amounts of tannins do not
succeed to expand weight and might ultimately die. The protective properties of tannins are usually characterized to their skill to unite proteins.
Protocatechllic and chlorogenic acids most likely have a special function
in disease resistance of certain plants. They check onion smudge, a disease
sourced by the fungus *Colletotrichum circinans,* and also deter germination of spore and growth of other fungi as well (Vickery & Vickery,
1981; Mayer, 1987). It is a deliberation by some that chlorogenic acid and
assured other related compounds can be readily formed and oxidized into
potent fungistatic quinones by certain disease-resistant cultivars but less
readily so by susceptible ones.

5.6.3 COUMARIN

Coumarins are simple phenolic compounds formed by a large diversity
of foliage in reaction to disease-causing organism or herbivore attack.
They are obtained from the shikimic acid pathway (Murray et al., 1982),
familiar in bacteria, fungi, and plants but missing in animals. In addition,
coumarins are also a highly active group of molecules with a broad series
of antimicrobial action against both fungi and bacteria (Brooker et al.,

2008). It is understood that these cyclic compounds perform as normal pesticidal defense compounds for plants and they signify an initial point for the examination of new derivatives possessing a series of better antifungal activity. Some coumarin imitative has high antifungal action against a series of soil-borne plant pathogenic fungi and shows more steadiness in contrast to the unique coumarin compounds alone (Brooker et al., 2008).

5.6.4   FURANOCOUMARINS

Furanocoumarin is a phenolic compound with particular concern of phytotoxicity and it is profuse in members of the family Umbelliferae with celery parsnip and parsley. Until they are stimulated by light (UV-A), these composites are not noxious; they might cause some furanocoumarins to develop into a stimulated high-power electronic state, which can add themselves into the double coil of DNA and attach to the pyramidine bases and thus blocking transcription and repair and ultimately leading to cell death (Rice, 1987). Psoralin, an essential linear furacoumarin, is recognized for its utilization in the action of fungal defense and observed very rarely in SO_2-treated plants (Ali et al., 2008).

5.6.5   FLAVONOIDS

The flavonoids contain a very large group within the phenolic compounds, while the anthocyanins, colorful in nature, are accountable for the impartation of the beautiful colors observed in the flowers, fruits, and leaves of plants known to produce water-soluble pigments that defend plants from the harmful effects of ultraviolet radiation (Mazid et al., 2011). In addition, two additional most important grouping of flavonoids observed in blossoms are flavones and flavonols, with a purpose to look after cells from UV-B radiation because they gather in epidermal layers of foliage and stems and absorb light sturdily in the UV-B region at the same time as letting visible photosynthetically active radiation (PAR) wavelengths all over unobstructed (Lake et al., 2009).

5.6.6 ISOFLAVONOIDS

Isoflavonoids are obtained from a flavonone intermediate, naringenin, all over in plants and perform a vital position in plant developmental and defense response. Though, phytoalexins are isoflavones with antibiotic and antifungal properties which are created in reaction to pathogen attack. These molecules are toxic that prevent disease-causing organism metabolism or cellular structure but are frequently pathogen specific in their harmful effect. Some examples of phytoalexin include medicarpin manufactured by alfalfa (*Medicago sativa*), rishitin generated by jointly tomatoes and potatoes (family: Solanaceae), and camalexin, created by *Arabidopsis thaliana*, reported by Freeman and Beattie (2008). Pterocarpans are also isoflavones that have been cutoff from the heartwood of *Pterocarpus macrocarpus* Kruz. with antifeedent activity against *Spodoptera litura* and *Reticulitermes speratus* (Morimoto et al., 2006). However, they also play a significant role in encouraging the improvement of nitrogen-fixing nodules by symbiotic rhizobia (Sreevidya et al., 2006). Moreover, it appears that creation of these flavonoids is an efficient strategy against ROS (Posmyk et al., 2009).

5.7 SULFUR-CONTAINING SECONDARY METABOLITES

These compounds include GSL, GSH, defensins, phytoalexins, thionins, and allinin which have been correlated directly or not directly with the defense of foliage in opposition to microbial pathogens (Crawford et al., 2000; Saito, 2004; Grubb & Abel, 2006; Halkier & Gershenzon, 2006).

5.7.1 GSL

Nitrogen and sulfur-holding plant glucosides having small molecular mass are produced by higher plants with the aim of enhancing their resistance against the unfavorable effects of biocontrol agents and competitors as their collapse products are discharged as volatile-protective substances showing toxic or repellent effects (Wallsgrove et al., 1999; DeVos & Jander, 2009), such as, mustard oil glucosides in Cruciferae and allyl-*cys*-sulfoxides in *Allium* (Leustek, 2002). The aroma volatiles from GSL, accelerated by myrosinase, split glucose from its bond with the S atom. The resultant aglycon reorganizes with loss of the sulfate to provide pungent and chemically reactive products, including isothiocyanates and nitriles, a role in defense as herbivorous poisons and feeding deterrent (Grubb & Abel, 2006; Halkier & Gershenzon, 2006; Talalay & Fahey, 2001). The strength of GSL arises when the foliage tissues are damaged and GSL comes into contact with the plant enzyme myrosinase which eliminates the β-glucose moiety leading to shape unbalanced intermediates, that is, isothiocyanates (R–N=C=S) and nitriles which in turn function in defense as herbivore poisons and feeding repellents (Geu-Flores et al., 2009; Ratzka et al., 2002; Zukalova & Vasak, 2002). They are metabolized and assimilated because isothiocyanates that may change the movement of enzymes are engaged both in the antioxidant defense system and in the detoxification from xenobiotics and considerably affect GST movement and cell defense against DNA injury (Lipka et al., 2010; Porrini, 2008), whereas toxicity of glucosinolatic products is well acknowledged but their mode of action has not so far been clarified and consequences from experimentations with *Brassica* plants modified in GSL content made doubts about their involvement in plant defense. Mithen and Magrath (1992) reported that the level of alkenyl GSL inside the foliage of a *Brassica* line that was resistant to *Leptosphearia maculans* was at all times repeatedly superior than that of a vulnerable line but they cannot find a correlation between the level of alkenyl GSL and disease resistance. Another report revealed the activity of numerous isothiocyanates against the nematode *Heterodera schachti* (Lazzeri et al,. 1993). Pedras and Sorensen (1998) found that the germination and radial growth pattern of spores of a potent Phoma lingam pathotype was deterred by higher concentrations of different isothiocyanates. A few studies of crosses of *Brassica* lines by diverse glucosinolate level show resistance to fungal injuries unsuccessful to compare with high and low glucosinolate level (Mithen & Magrath, 1992; Wretblad & Dixelius, 2000).

5.7.2   GSH

This compound is the vital form of organic S in the soluble portion of plants and has a key role as a mobile pool of reduced S in the regulation of plant expansion and growth, and as a cellular antioxidant in stress reactions (Kang and Kim, 2007; Noctor and Foyer, 1998). Furthermore, GSH is also concerned in the detoxification of xenobiotics and cytotoxins by aiming them in to vacuole (Rea et al., 1998). It is quickly built up after fungal attack, and may perform as systemic messenger carrying information involving the attack to non-infested tissues.

5.7.3 PHYTOALEXINS

They are made in reaction to bacterial or fungal disease or further forms of strain and aid in restraining the attack pathogens by gathering around the site of infection. This has emerged to be a familiar method of resistance to pathogenic microbes in a large series of plants (Van Etten et al., 1994; Grayer and Harborne, 1994). A great number of the plant families manufacture organic phytoalexins of diverse chemistry; these groups are frequently linked with a family, for example, isoflavonoids of Leguminosae, sesquiterpenoids of Solanaceae, while phytoalexins from *Brassica* have an indole or connected ring system and one sulfur atom as regular structural features. Simultaneously, the Crucifereae shows to be the only plant family manufacturing these S metabolites (Gross et al., 1993; Pedras and Sorensen, 1998), which are obviously dissimilar from the other recognized GSL (Harborne, 1999).

5.7.4 DEFENSINS, THIONINS, AND LECTINS

These all compounds are S-loaded non-storage plant proteins that produce and gather after a microbial injury and such related situations (Van Loon et al., 1994). These all restrict the development of a wide range of fungi (Thomma et al., 2002). A few defensins are antifungal or occasionally show antibacterial activity (Thomma et al., 2002). Furthermore defensins genes are partly pathogen-inducible (Gu et al., 1992) and others that are involved in resistance can be expressed constitutively (Parashina et al., 2000). The components appear to be mixed up in the natural defense method of plants as they can be extremely toxic to microorganisms, insects, and mammals.

Accumulation of thionins inside the cell wall of infected wheat spikes of resistant wheat cultivars signifies that the development of thionins might be implicated in defense reactions to infections and in dispersion of *Fusarium culmorum* (Kang & Buchenauer, 2003). A few plant species manufacture lectins as defensive proteins that fasten to carbohydrate or carbohydrate-containing proteins. After being ingested by herbivorous insects, lectins fasten to epithelial cell lining of the digestive tracts and obstruct with nutrient incorporation (Peumans & Van Damme, 1995).

5.8 BACKGROUND AND PRESENT SCENARIO OF THE IMPROVEMENT AND EXPLOITATION OF PLANT METABOLITES IN PEST MANAGEMENT

In the past, cultivational controls such as collection of resistant crops, rotational farming, and physical controls were utilized for crop protection. Steadily, diverse chemicals with pest control properties were exploited in artificial ecosystems. Plants having antimicrobial properties and their metabolites have been utilized by mankind since ancient times, particularly in cultures with strong herbal tradition (Singh & Anuradha, 2011). The Chinese people exploited *Pyrethrum* and *Derris* species and the Romans hellebore species as insect management agents. During early times of agricultural expansion, spices, for example, cinnamon, mustard, nutmeg, pepper, were utilized to protect food from insect attack (Isman, 2006). However, before the Second World War, four most important groups of composites were usually used: nicotine and alkaloids, rotenone and rotenoids, pyrethrum and pyrethrins, and vegetable oils (Isman, 2006).

A few of them had several awkward properties because of their toxicity on nontarget species (nicotine) or the unsteadiness of the molecules (pyrethrum). Therefore, the utility of these matters decreased with the commercial availability of chemically manufactured insecticides which moreover were easier to synthesize and handle and were less costly (Regnault-Roger, 2012). The exploitation of botanical pesticides was referred to markets, such as household products, garden, and veterinary purposes. Current history illustrates that the incessant and immense use of synthetic pesticides has produced numerous surprising side effects, such as acute and chronic toxicity to human, expansion of resistance in pests, removal of natural biocontrol and pollination agents, insect resurgence, effects on nontarget organisms, and ecological contamination with efficient effects on the entire food chain (Isman, 2006; Akhtar et al., 2009; Koul et al., 2008). Governments reacted to these problems with regulatory action, prohibition, or severely restricting the most damaging products, and generating policies to substitute synthetic pesticides of concern with those verified to present minor risks to human health and the atmosphere (Isman, 2006; Regnault-Roger, 2012). Therefore, the regulation of plant protection commodities in the European Union (EU), persuading the natural chemical pesticides, was first harmonized under Directive 91/414/EEC in 1993. This directive established settled criterion for allowing the safety of active substances, with the safety and effectiveness of formulated products. In 2011, Regulation (EC) 1107/2009 initiated some new criteria for registration of plant protection commodities from plant origin as basic materials and low-risk pesticides. These alterations in the regulatory environment appeared to intensify the momentum for the discovery and improvement of alternative pest management products—those with diminished strength and environmental impacts—including pesticides resulting from plants (Isman, 2006). The scientific literature depicts hundreds of segregated secondary metabolites that show behavior (repellence, oviposition deterrence, feeding deterrence) or physiologic (acute toxicity, developmental disruption, growth inhibition) consequences to pests in any case in laboratory bioassays (Fischer et al., 2013). Even though so many products have been isolated, differentiated, and evaluated as pesticidal compounds, not much advancement has been made in the commercial availability of such products. The greater part of this text deals with compounds with potential activities that are not commercially existing (Copping & Duke, 2007). In spite of the level of this research endeavor, only a few of plant-based insecticides are in commercial use on vegetable and fruit crops today, with considerable marketable progress of only two new sources of botanicals in the past 20 years (neem and essential oils) (Isman, 2006). For decades, the use of botanicals was more focused on the management of insects than on other plant pests; they are repellent, antifeedant, ovipositional deterrent, and/or neurotoxic. More in general, they affect the biotic potential of parasites and pests. Plant meatabolites and allelochemicals also perform on a broad range of species like nematodes, phytopathogenic microorganisms, with other plant species (allelopathy). In current years, the expansion of information on plant resistance mechanisms against bio-aggressors highlighted that plant allelochemicals perform an essential role in plant defense and some botanicals were recognized as inducers of resistance and already recorded and marketed (Regnault-Roger, 2012). Some of the plant natural compounds and their active components, mode of action, and commercial application as crop protection agents are given in Table 5.2.

However, major obstacles to commercialization of plant-based pesticides are (a) accessibility and sustainability of the botanical resource; (b) strength, standardization, and quality control of the chemically composite extracts based on quantification of active ingredients; and (c) the regulatory support, usually requiring expensive toxicological estimation of the candidate product (Isman & Paluch, 2011; Fischer et al., 2013). Regarding the accessibility of biomass and sustainable utilization of the plant species, resources that can be established as a crop (if possible during all the year) are recommended to guarantee the delivery of raw material for manufacturing botanical pesticides on a commercial scale, unless it has an enormously high natural abundance. Furthermore, the standardization of natural product-based anti-insect preparations has really the biggest constraint and these preparations have subsequently been hindered in their potential marketability compared with usual pesticides (Fischer et al., 2013). For a botanical insecticide to give a consistent level of effectiveness to the user, there should be some point of chemical consistency, most likely based on the putative active ingredient(s). For achieving standardization, the manufacturer should have a diagnostic method and the equipment essential for analysis and might need to mix or blend extracts from diverse sources, which needs storage facilities and is partly dependent on the inherent steadiness of the active principles in the source plant material or extracts thereof held in storage (Isman, 2006).

However, under reflections based on evaluations and reliable data, many natural products are now considered to be pesticides of least threat (Regnault-Roger, 2012). For a few plants and plant extracts, which are recently recorded in «25b list» of the US EPA and all matters with GRAS status, reduced numbers are needed for the registration (Cavoski et al., 2011). Regnault-Roger (2012) reported that the worldwide market for biopesticides was of a value of an estimated $1.6 billion in 2009, but is projected to enhance to $3.3 billion in 2014 for a 5-year compound annual growth rate of 15.6%. These numbers highlight that biopesticide industry expansion continues, but improvement has to be done before biopesticides share the plant protection products market with synthetic pesticides (biopesticide market symbolized 2.5% of the global pesticide market in 2005). North America is leading in the use of biocontrol products (44%), followed by Asia (24%) and Europe (20%); Africa and South-Central America represent only 14%. However, the demand for natural biopesticides is getting higher progressively in all parts of the planet.

TABLE 5.2 Some Commercial Natural Plant Products, Their Active Compound and Mode of Action.

Name of the plant product	Plant source	Active compounds	Biological activity	Mode of action	References
Neem (neem oil, medium polarity extracts)	*Azadirachta indica* A. Juss	Azadirachtin, dihydroazadirachtin, variety of triterpenoids (nimbin, salannin, etc.)	Insecticide, acaricide, fungicide	Molting inhibitors (ecdysone antagonists), antifeedant/repellent	Isman (2006), Copping and Duke (2007), Isman and Paluch (2011)
Pyrethrum	*Tanacetum cinerariaefolium* Schultz-Bip.	Esters of chrysanthemic acid and pyrethric acid (pyrethrins I and II, cinerins I and II, jasmolins I and II)	Insecticide, acaricide	Axonic poisons (sodium channels, agonists)	Isman (2006), Copping and Duke (2007), Isman and Paluch (2011)
Nicotine	*Nicotiana* spp.	(S)-isomer, (RS)-isomers, and (S)-isomer of nicotine sulfate	Insecticide	Neurotoxin (acetylcholine agonist)	Isman (2006),Copping and Duke (2007), Isman and Paluch (2011)
Rotenone	*Derris, Lonchocarpus* and *Tephrosia* species	Rotenone, deguelin, (isoflavonoids)	Insecticide, acaricide	Mitochondrial cytotoxin	Isman (2006), Copping and Duke (2007), Isman and Paluch (2011)
Ryania	*Ryania* spp. (*Ryania speciosa* Vahl)	Ryanodine, ryania, 9,21-didehydroryanodine (alkaloids)	Insecticide	Neuromuscular poison (calcium channel agonist)	Isman (2006), Copping and Duke (2007), Isman and Paluch (2011)
Sabadilla	*Schoenocaulon* spp. (*Schoenocaulon officinale* Gray)	Mixture of alkaloids (cevadine, veratridine)	Insecticide	Axonic poisons (sodium channels agonists, heart and skeletal muscle cell membranes)	Isman (2006), Copping and Duke (2007), Isman and Paluch (2011)
Quassia	Quassia, Aeschrion, Picrasma	Quassin (triterpene lactone)	Insecticide	Unknown	Isman (2006), Isman and Paluch (2011)

TABLE 5.3 *(Continued)*

Name of the plant product	Plant source	Active compounds	Biological activity	Mode of action	References
Extract of giant knotweed	*Reynoutria sachalinensis* (Fr. Schm.) Nakai	Physcion, emodin	Fungicide, bactericide	Induction of SAR (phenolic phytoalexines)	Dayan et al. (2009), Regnault-Roger (2012)
Karanjin	*Derris indica* (Lam.) Bennet	Karanjin	Insecticide, acaricide	Antifeedant/repellent, insect growth regulator	Copping and Duke (2007)
Clove essential oil	*Syzygium aromaticum*, *Eugenia caryophyllus* Spreng	Eugenol (mixture of several predominantly terpenoid compounds)	Insecticide, herbicide	Neurotoxic, interference with the neuromodulator octopamine	Isman (2006), Copping and Duke (2007), Regnault-Roger (2012)
Cinnamon essential oil	*Cinnamomum zeylanicum*	Cinnamaldehyde	Insecticide, herbicide	Octopamine antagonists; membrane disruptors, others	Dayan et al. (2009)
Lemon grass essential oil	*Cymbopogon nardus*, *Cymbopogon citratus* Stapf., *Cymbopogon flexuosus* D.C.	Citronellal, citral	Insecticide, herbicide	Octopamine antagonists; membrane disruptors, others	Dayan et al. (2009)
Mint essential oil	*Mentha* species (mint)	Menthol	Insecticide	Octopamine antagonists; membrane disruptors, others	Dayan et al. (2009)
Citronella oil	*Cymbopogon* spp.	Citronellal, geraniol, other terpenes	Repellent, herbicide		Dayan et al. (2009)

Nowadays, interest in commercially available natural pest management materials for use in organic agriculture and "green" pest management has grown substantially (Copping & Duke, 2007). This strong increase matches with the progress of control of insect pest in the sector of high-value crops like vegetables in greenhouses, vineyard, tree, and fruit farming. Biocontrol will expand not only through the enhancement of organic farming but also through Integrated Pest Management (Regnault-Roger, 2012). Also, due to their less residual effect and mammalian toxicity, these products may be functional indoor, in backyards, in treatments small time before and post harvest and their effect will protect crop products during transportation and storage with minimum risk for consumers.

In developing countries, the conventional use of foliage and their derivatives for defense of stored products is long established. The potency of secondary metabolites for plant protection could be employed in a more recent alternative strategy aiming at strengthening the plant defense by rising its own mechanisms through allelochemicals (Regnault-Roger, 2012). These products can be described on the basis that different commodities in the mixture might have diverse modes of action or target sites in the insect or are able to restrict the detoxification enzymes that normally degrade a single constituent (Akhtar et al., 2009).

The push–pull strategy is an example of a harmonized control strategy involving the behavioral management of insects. Push–pull suggests the use of groupings of deterrents/repellents and attractive stimuli to direct the movement of insects away from protected resources. Economic crops are confined from pests by repellent plants, ovipositional deterrents or anti-feedants. Simultaneously, pests are restricted on trap crops, using aggregative semichemicals and attractants, so that a selective management (e.g., biological control) can be used to decrease pest populations (Moreau & Isman, 2012).

5.9 CONCLUSION

Secondary plant metabolites from plant tissues are a significant source of biopesticides and the progress of antifeedants. Plants have progressed multiple defense mechanisms against microbial agents and various types of environmental stress. A few antimicrobial secondary metabolites are executed and induced by pathogenic infection. At present, research activities in different parts of the world are focusing attention on possible utilization of higher plant products such as newer chemotherapeutics for plant protection because they are generally non-phytotoxic and easily biodegradable (Isman, 2006). It has also been believed that some secondary metabolites cutoff from a plant that has been exposed to have no bioactivities or antifeedant properties against insect pests can be utilized for their potentials by mixing these isolated constituents with some insecticide.

At present, different plant based-products have been prepared for huge-scale application as biopesticides and antifeedants in ecological management of foliage pests and are being utilized as substitutes to synthetic pesticides in crop protection and post-harvest usage. These products have lower mammalian poisons and are cost-effective. Such products of elevated plant origin may be utilized as ecochemicals and integrated into plant protection programs. Current reviews of the key chemical classes of plant metabolites which have been exploited in crop protection and post-harvest focused on the most recent advances in the chemicals revealed their mode of work and their destiny in the environment, their current use in pest management, related registration trials, and commercialization potential. Botanicals are included in Integrated Pest Management (IPM) programs and result in crop protection.

There is a broad scale for the utilization of botanical pesticides in the IPM of diverse insect pests. Production of botanical antifeedants would reduce the maximum cost of importation in emerging countries.

KEYWORDS

- secondary metabolites
- defense mechanism
- management
- pest
- abiotic stress

REFERENCES

Akhtar, Y.; Shikano, I.; Isman, M. B. Topical Application of a Plant Extract to Different Life Stages of *Trichoplusia ni* Fails to Influence Feeding or Oviposition Behaviour. *Entomol. Experim. Appl.* **2009**, *132*, 275–282.

Athanasiadou, S.; Kyriazakis, I. Plant Secondary Metabolites: Antiparasitic Effects and Their Role in Ruminant Production System. *Proc. Nutr. Soc.* **2004**, *63*, 631–639.

Ali, S. T.; Mahmooduzzafar-Abdin, M. Z.; Iqbal, M. Ontogenetic Changes in Foliar Features and Psoralen Content of *Psoralea corylifolia* Linn. Exposed to SO_2 Stress. *J. Environ. Biol.* **2008**, *29*(5), 661–668.

Banchio, E.; Valladares, G.; Defago, M.; Palacios, S. Y.; Carpinella, C. Effects of Melia *azedarach* (Meliaceae) Fruit Extracts on the Leafminer *Liriomiza huidobrensis* (Diptera, Agromyzidae): Assessment in Laboratory and Field Experiments. *Ann. Appl. Biol.* **2003**, *143*, 187–193.

Baysal, O. Determination of Microorganisms Decomposing Essential Oils of *Thymbra spicata* L. var. *spicata* and Effect of these Microorganisms on Some Soil Borne Pathogens. M. Sc. Thesis, Akdenniz University, Antalya, 1997. In Dubey, N. K., Srivastava, B., and Kumar, A. "Current Status of Plant Products as Botanical Pesticides in Storage Pest Management". *J. Biopesticides* **2008**, *1*(2), 182–186.

Belles, X.; Camps, F.; Coll, J.; Dollars, P. M. Insect Antifeedant Activity of Clerodane Diterpenoids against Larvae of *Spodoptera littoralis* (Lepidoptera). *J. Chem. Ecol.* **1985**, *11*, 1439–1445.

Bobadilla, M.; Zavala, F.; Sisniegas, M., Zavaleta, G.; Mostacero, J. Y.; Taramona, L. Evaluación Larvicida de Suspensiones Acuosas de *Annona muricata* Linnaeus (guanábana) sobre *Aedes aegypti* Linnaeus (Diptera, Culicidae). *Rev. Peruana Biol.* **2005**, *12*, 15–152.

Bobadilla, M.; Zavaleta, G.; Gil, F.; Pollack. L. Y.; Sisniegas, M. Efecto Bioinsecticida Del Extracto Etanólico de Las Semillas de *Annona cherimolia* Miller (chirimoya) y *A. muricata* Linnaeus (guanábana) Sobre Larvas del IV Estadio de *Anopheles* sp. *Rev. Peruana Biol.* **2002**, *9*, 64–73.

Bradley, D. J.; Kjellborn, P.; Lamb, C. J. Elicitor and Wound Induced Oxidative Cross Linking of a Proline Rich Plant Cell Protein: A Novel Rapid Defence Response. *Cell* **1992**, *70*, 21–30.

Brooker, N.; Windorski, J.; Blumi, E. Halogenated Coumarins Derivatives as Novel Seed Protectants. *Commun. Agric. Appl. Biol. Sci.* **2008**, *73*(2), 81–89.

Camps, F.; Coll, J.; Cortel, A. Two New Clerodane Diterpenoids from *Ajuga reptans* (Labiatae). *Chem. Lett.* **1981**, *8*, 1093–1096.

Camps, F.; Coll, J.; Cortel, A. New Clerodane Diterpenoids from *Ajuga iva*. *Chem. Lett.* **1982**, *9*, 1053–1056.

Camps, F.; Coll, J. Insect Allelochemicals from Ajuga Plants. *Phytochemistry* **1993**, *32*, 1361–1370.

Cavalcante, G.; Carrano, A. Y.; Dias, S. Potencialidade Inseticida de Extratos Aquosos de Esencias Florestais Sobre Mosca Branca. *Pesq. Agropec. Bras.* **2006**, *41*, 9–14.

Cavoski, I.; Caboni, P.; Miano, T. Natural Pesticides and Future Perspectives. In *Pesticides in the Modern World—Pesticides Use and Management*; Stoytcheva, M.,

Ed.; 2011, pp 169–190. Available from http://www.intechopen.com/articles/show/title/ naturalpesticides-and-future-perspectives.

Celis, A.; Mendoza, C.; Pachón, M.; Cardona, J.; Delgado, W. Y.; Cuca, L. Extractos Vegetales Utilizados como Biocontroladores con énfasis en la familia Piperácea. Una Revisión. *Agron. Colomb.* **2008**, *26*, 97–106.

Cole, M. D.; Anderson, J. C.; Blaney, W. M.; Fellows, L. E.; Ley, S. V.; Sheppard, R. N. Neoclerodane Insect Antifeedants from *Scutellaria galericulata*. *Phytochemistry* **1990**, *29*, 1793–1796.

Copping, L. G.; Duke, S .O. Natural Products that Have Been Used Commercially as Crop Protection Agents. *Pest Manage. Sci.* **2007**, *63*, 524–554.

Crawford, N. M.; Kahn, M. L.; Leustek, T.; Long, S. R. Nitrogen and Sulfur. In *Biochemistry and Molecular Biology of Plants*; Buchanan, B. B., Gruissem, W., Jones, R. L., Eds.; American Society of Plant Biologists: Rockville, MD, **2000**, pp 824–849.

Croteau, R.; Kutchan, T. M.; Lewis, N. G. Natural Products (Secondary Metabolites). *Biochem. Mol. Biol. Plants* **2000**, 1250–1318.

Davies, P. J. The Plant Hormone Concept: Concentration, Sensitivity and Transport. In *Plant Hormones: Physiology, Biochemistry and Molecular Biology*; Davies, P. J., Ed.; Kluwer, Boston, MA, 1995; pp 13–38.

Dayan, F. E.; Cantrell, C. h. L.; Duke, S. O. Natural Products in Crop Protection. *Bioorganic Med. Chem.* **2009**, *17*, 4022–4034.

DeVos, M.; Jander, G.. *Myzus persicae* (Green Peach Aphid) Salivary Components Induce Defence Responses in *Arabidopsis thaliana*. *Plant Cell Environ.* **2009**, *32*(11), 1548–1560.

Eigenbrode, S. D.; Thumble, J. T.; Millar, J. G.; White, K. K. Topical Toxicity of Tomato Sesquiterpenes to the Beat Armyworm and the Role of these Compounds in Resistance Derived from an Accession of *Lycopersicon hirsutum* f. *typicum*. *J. Agric. Food Chem.* **1994**, *42*, 807–810.

Eisner, T.; Meinwald, J. Chemical Ecology: The Chemistry of Biotic Interaction. National Academy Press: Washington, DC, 1995.

FAO. *Pesticide Residues in Food. Report*, vol. *116*, 1992, p 146.

Fischer, D.; Imholt, C.; Pelz, H. J.; Wink, M.; Prokopc, A.; Jacoba, J. The Repelling Effect of Plant Secondary Metabolites on Water Voles, *Arvicola amphibious*. *Pest. Manage. Sci.* **2013**, *69*, 437–443.

Freeman, B. C.; Beattie, G. A. An Overview of Plant Defenses against Pathogens and Herbivores. *Plant Health Instruct.* **2008**, *226*(1).

Garcia-Mateos, R.; Pérez, P. R.; Rodríguez, H. C. Y.; Soto, H. M.; Toxicidad De Alcaloides De *Eythrina americana* en Larvas De Mosquito *Culex quinquefasciatus*. *Fitotec. Mex.* **2004**, *27*, 297–303.

Gershenzon, J.; Croteau, R. Terpenoids. In *Herbivores Their Interaction with Secondary Plant Metabolites, Vol I: The Chemical Participants*, 2nd ed., Rosenthal, G. A., Berenbaum, M. R., Eds.; Academic Press: San Diego, **1991**; pp 165–219.

Geu-Flores, F.; Olsen, C. E.; Halkier, B. A. Towards Engineering Glucosinolates into Non-cruciferous Plants. *Planta* **2009**, *229*(2), 261–270.

Gonzalez-Coloma, A.; Gutierrez, C.; Del Corral, J. M. M.; Gordaliza, M.; De La Puente, M. L.; San Feliciano, A. Structure- and Species-dependent Insecticidal Effects of Neo-clerodane Diterpenes. *J. Agric. Food Chem.* **2000**, *48*, 3677–3681.

Gould, J. M. Probing the Structure and Dynamics of Lignin In Situ. *What's New Plant Physiol.* **1983**, *14*, 25–91.

Grayer, R. J.; Harborne, J. B. A Survey of Antifungal Compounds from Higher Plants 1982–1993. *Phytochemistry* **1994**, *37*, 19–42.

Gross, D.; Porzel, A.; Schmidt, J. Phytoalexine mit Indolstruktur aus Kohlrabi [Phytoalexins with indole structure from Kohlrabi]. *Zeitsch. Naturforsch.* **1994**, *49*(5–6), 281–285.

Grubb, C.; Abel, S. Glucosinolate Metabolism and its Control. *Trends Plant Sci.* **2006**, *11*, 89–100.

Gu, Q.; Kawata, E. E.; Morse, M. J.; Wu, H. M.; Cheung, A. Y. A Flower Specific CDNA Encoding a Novel Thionin in Tobacco. *Mol. Gen. Genet.* **1992**, *234*, 89–96.

Gupta, V. N.; Datta, S. K. Influence of Gibberellic Acid on Growth and Flowering in *Chrysanthemum* (*Chrysanthemum morifolium* rahmat) cv. Jayanti. *Indian J. Plant Physiol.* **2001**, *6*, 420–422.

Harborne, J. B. The Comparative Biochemistry of Phytoalexin Induction in Plants. *Biochem. Syst. Ecol.* **1999**, *27*, 335–367.

Halkier, B. A.; Gershenzon, J. Biology and Biochemistry of Glucosinolates. *Annu. Rev. Plant Biol.* **2006**, *57*, 303–333.

Hassan, A. M. M. A Review of Secondary Metabolites from Plant Materials for Post Harvest Storage. *Int. J. Pure Appl. Sci. Technol.* **2011**, *6*(2), 94–102.

Hassan, A. M. M. The Potential of Secondary Metabolites in Plant Material as Deterents against Insect Pests: A Review. *Afr. J. Pure Appl. Chem.* **2010**, *4*(11), 243–246.

Hough-Goldstein, J. A. Antifeedant Effects of Common Herbs on the Colorado Potato Beetle (Coleoptera: Chrysomelidae). *Environ Entomol.* **1990**, *19*, 234–238.

Isman, M. B. Botanical Insecticides, Deterrents, and Repellents in Modern Agriculture and an Increasingly Regulated World. *Annu. Rev. Entomol.* **2006**, *51*, 45–66.

Isman, M. B.; Paluch, G. Needles in the Haystack: Exploring Chemical Diversity of Botanical Insecticides. In *Green Trends in Insect Control*; López, O., Fernández-Bolaños, J. G., Eds.; RSC, 2011; pp 248–265.

Ivie, G. W.; Witzel, D. A. Sesquiterpene Lactones: Structure, Biological Action and Toxicological Significance. In *Handbook of Natural Toxins*; Keeler R. F., Tu, A. T., Eds.; Marcel Dekker Inc.: New York, 1982; pp 543–584.

Jansen, B. J. M.; Groot, A. The Occurrence and Biological Activity of Drimane Sesquiterpenoids. *Nat. Prod. Rep.* **1991**, *8*, 309–318.

Kang, S. Y.; Kim, Y. C. Decursinol and Decursin Protect Primary Cultured rat Cortical Cells from Glutamate-Induced Neurotoxicity. *J. Pharm. Pharmacol.* **2007**, *59*(6), 863–870.

Kang, Z.; Buchenauer, H. Immonocytochemical Localization of Cell Wall-Bound Thionins and Hydroxyproline-Rich Glycoproteins in *Fusarium culmorum*-Infected Wheat Spikes. *J. Phytopathol.* **2003**, *151*(3), 120–129.

Karunamoorthi, K.; Ramanujam, S. Y.; Rathinasamy, R. Evaluation of Leaf Extracts of *Vitex negundo* L. (Family: Verbenaceae) against Larvae of *Culex tritaeniorhynchus* and Repellent Activity on Adult Vector Mosquitoes. *Parasitol. Res.* **2008**, *103*, 545–550.

Khan, N. A.; Singh, S. *Abiotic Stress and Plant Responses.* I. K. Int. Publ. House Pvt. Ltd.: S-25, Green Park Extension, Uphaar Cinema Market, New Delhi-110016 India, 2008.

Klein, R. M. *The Green World: An Introduction to Plants and People.* Harper and Row: New York, 1987.

Koul, O. *Insect Antifeedants*. CRC Press: Boca Raton, FL, 2005.

Koul, O. Biological Activity of Volatile D-*n*-Propyl Difulside from Seeds of Neem, *Azadirachta indica* (Meliaceae), to Two Species of Stored Grain Pests, *Sitophilus oryzae* (L.) and *Tribolium oryzae* (Herbst). *J. Econ. Entomol.* **2004**, *97*, 1142–1147.

Koul, O.; Walia, S.; Dhaliwal, G. S.; Essential Oils as Green Pesticides: Potential and Constraints. *Biopest. Int.* **2008**, *4*, 63–84.

Kubo, I.; Klocke, J. A.; Miura, I.; Fukuyama, Y. Structure of Ajugarin-IV. *J. Chem. Soc. Chem. Commun.* **1982**, *11*, 618–619.

Kubo, I.; Nakanishi, K. Some Terpenoid Insect Antifeedants from Tropical Plants. In *Advances in Pesticide Science, Part 2*; Geissbuhler H., Ed.; Pergamon Press: Oxford, 1978; pp 284–294.

Kubo, I.; Fukuyama, Y.; Chapya, A. Structure of Ajugarin-IV. *Chem. Lett.* **1983**, *10*, 223.

Lajide, L.; Escoubas, P.; Mizutani, J. Termite Antifeedant Activity in *Detarium microcarpum*. *Phytochemistry* **1995a**, *40*, 1101–1104.

Lajide, L.; Escoubas, P.; Mizutani, J. Termite Antifeedant Activity in *Xylopia aethiopica*. *Phytochemistry* **1995b**, *40*, 1105–1112.

Lake, J. A.; Field, K. J.; Davey, M. P.; Beerling, D. J.; Lomax, B. H. Metabolomic and Physiological Responses Reveal Multi-Phasic Acclimation of *Arabidopsis thaliana* to Chronic UV Radiation. *Plant Cell Environ.* **2009**, *32*(10), 1377–1389.

Lazzeri, L.; Tacconi, R.; Palmieri, S. In Vitro Activity of Some Glucosinolates and their Reaction Products toward a Population of the Nematode *Heterodera schachtii*. *J. Agric. Food Chem.* **1993**, *41*, 825–829.

Leustek, T. Sulfate metabolism. In *The Arabidopsis Book*; Somerville, C. R., Meyerowitz, E. M. Eds.; American Society of Plant Biologists: Rockville, MD. DOI:10.1199/tab.0009.2002.

Lewis, N. G.; Yamamoto, E. Lignin: Occurrence, Biogenesis and Biodegradation. *Ann. Rev. Plant Phys. Plant Molecular Biol.* **1990**, *41*, 455–496.

Lewis, W. H.; Elvin-Lewis M. P. F. Medical Botany: Plants Affecting Mans Health. Wiley: New York, 1977.

Lipka, U.; Fuchs, R.; Kuhns, C.; Petutschnig, E.; Lipka, V. Live and Let die-Arabidopsis Non-host Resistance to Powdery Mildews. *Eur. J. Cell Biol.* **2010**, *89*(2), 194–199.

Lwande, W.; Hassanali, A.; Mcdowell, P. G.; Moreka, L.; Nokoe, S. K; Waterman, P. G. Constituents of *Commiphora rostrata* and some of their Analogues as Maize Weevil, *Sitophilus zeamais* Repellents. *Insect Sci. Appl.* **1992**, *13*, 679–683.

Macas, F. A.; Simonet, A. M.; Galindo, J. C. G.; Castellano, D. Bioactive Phenolics and Polar Compounds from *Meliolotus messanensis*. *Phytochemistry* **1999**, *50*, 35–46.

Mader, M.; Amberg-Fisher, V. Role of Peroxidase in Lignifications of Tobacco Cells. Oxidation of Nicotinimide Adenine Dinucleotide and Formation of Hydrogen Peroxide by Cell Wall Peroxidises. *Plant Physiol.* **1982**, *70*, 1128–1131.

Mauricio, R. Costs of Resistance to Natural Enemies in Field Populations of the Annual Plant *Arabidopsis thaliana*. *Am. Nat.* **1998**, *151*(1), 20–28.

Mayer, A. M. Polyphenols Oxidase in Plants—Recent Progress. *Phytochemistry* **1987**, *26*, 11–20.

Mazid, M.; Khan, T. A.; Mohammad, F. Role of Secondary Metabolites in Defense Mechanisms of Plants. *Biol. Med.* **2011**, *3*(2), 232–249.

Mithen, R.; Magrath, R. Glucosinolates and Resistance to *Leptosphaeria maculans* in Wild and Cultivated *Brassica* Species. *Plant Breed.* **1992**, *108*, 60–68.

Modupe, M. A.; Musa, M. Prospect of Antifeedant Secondary Metabolites as Post Harvest Material. *Int. J. Innov. Res. Sci. Eng. Technol.* **2014**, *3*(1), 8701–8708.

Mordue, A. J.; Blackwell, A. Azadirachtin: An update. *J. Insect Physiol.* **1993**, *39*, 903–924.

Moreau, T. L.; Isman, M. B. Combining Reduced-Risk Products, Trap Crops and Yellow Sticky Traps for Greenhouse Whitefly (*Trialeurodes vaporariorum*) Management on Sweet Peppers (*Capsicum annum*). *Crop Prot.* **2012**, *34*, 42–46.

Morimoto, M.; Fukumoto, H.; Hiratani, M.; Chavasir, W.; Komai, K. Insect Antifeedants, Pterocarpans and Pterocarpol in Heartwood of *Pterocarpus macrocarpus* Kruz. *Biosci. Biotechnol. Biochem.* **2006**, *70*(8), 1864–1868.

Munoz, D. M.; Dela Torre, M. C.; Rodriguez, B.; Simmonds, M. S. J.; Blaney, W. M. Neo-Clerodane Insect Antifeedants from *Scutellaria alpina* sub sp. *Javalambrensis*. *Phytochemistry* **1997**, *44*, 593–597.

Murray, R. D. H.; Mendez, J.; Brown, S. A. *The Natural Coumarins*. Wiley: New York, 1982.

Ndungu, M.; Lwande, W.; Hassanali, A.; Moreka, L.; Chhabra, S. C. *Cleome monophylla* Essential oil and its Constituents as Tick *Rhipicephalus appendiculatus* and Maize weevil (*Sitophilus zeamais*) Repellents". *Entomol. Exp. Appl.* **1995**, *76*, 217–222.

Noctor, G.; Foyer, C. H. Ascorbate and Glutathione: Keeping Active Oxygen under Control. **1998**.

Omar, S.; Marcotte, M.; Fields, P.; Sanchez, P. E.; Poveda, L.; Mata, R. Antifeedant Activities of Terpenoids Isolated from Tropical Rutales. *J. Stored Prod. Res.* **2007**, *43*, 92–96.

Ortego, F.; Rodriguez, B.; Castanera, P. Effects of Neo-Clerodane Diterpenes from *Teucrium* on Feeding Behavior of Colorado Potato Beetle Larvae. *J. Chem. Ecol.* **1995**, *21*, 1375–1386.

Owen, T. Geoponika: Agricultural Pursuits. In *An Overview of Plants as Insect Repellents*; Moore, S. J., Langlet, A., Eds, 2004. In *Traditional Medicine, Medicinal Plants and Malaria*; Wilcox, M., Bodeker, G., Eds; Taylor and Francis, London, 2004. http://www. ancientlibrary.com/geoponica/index.html, 1805.

Oxenham, S. K. Classification of an *Ocimum basilicum* Germplasm Collection and Examination of the Antifungal Effects of the Essential oil of Basil. Ph. D. Thesis, University of Glasgow, Glasgow, UK, 2003; In: Dubey, N. K.; Srivastava, B.; Kumar, A. Current Status of Plant Products as Botanical Pesticides in Storage Pest Management. *J. Biopest.* **2008**, *1*(2), 182–186.

Parashina, E. V.; Serdobinskii, L. A.; Kalle, E. G.; Lavorova, N. A.; Avetisov, V. A.; Lunin, V. G.; Naroditskii, B. S. Genetic Engineering of Oilseed Rape and Tomato Plants Expressing a Radish Defensin Gene. *Russ. J. Plant Physiol.* **2000**, *47*, 417–423.

Pedras M. S. C.; Sorensen J. L. Phytoalexin Accumulation and Antifungal Compounds from the Crucifer wasabi. *Phytochemistry* **1998**, *49*, 1959–1965.

Perez-Pacheco, R.; Rodríguez, C.; Lara, J.; Montes, R. Y.; Valverde, G. Toxicidad de aceites, esencias y extractos vegetales en larvas de mosquito *Culex quinquefasciatus* Say (Diptera: Culicidae). *Acta Zool. Mex.* **2004**, *20*, 141–152.

Peumans, W. J.; Van Damme, E. J. M. Lectins as Plant Defence Proteins. *Plant Physiol.* **1995**, *109*, 347–342.

Pichersky, E.; Gershenzon, J. The Formation and Function of Plant Volatiles: Perfumes for Pollinator Attraction and Defense. *Curr. Opin. Plant Biol.* **2002**, *5*, 237–243.

Picman, A. K. Biological Activities of Sesquiterpene Lactones. *Biochem. Syst. Ecol.* **1986**, *14*, 255–281.

Porrini, M.; Functional Foods: From Theory to Practice. *Int. J. Vitamin Nutr. Res.* **2008**, *78*(6), 261–268.

Posmyk, M. M.; Kontek, R.; Janas, K. M. Antioxidant Enzymes Activity and Phenolic Compounds Content in Red Cabbage Seedlings Exposed to Copper Stress. *Ecotoxicol. Environ. Saf.* **2009**, *72*(2), 596–602.

Ratzka, A.; Vogel, H.; Kliebenstein, D. J.; Mitchell-Olds, T.; Kroymann, J. Disarming the Mustard Oil Bomb. *Proc. Natl. Acad. Sci.* **2002**, *99*(17), 11223–11228.

Rea, P. A.; Li, Z. S.; Lu, Y. P.; Drozdowicz, Y. M.; Martinoia, E. From Vacuolar GS-X pumps to Multispecific ABC Transporters. *Ann. Rev. Plant Physiol. Plant Mol. Biol.* **1998**, *49*, 727–760.

Regnault-Roger, C. Trends for Commercialisation of Biocontrol Agent (Biopesticide) Products. In *Plant Defence: Biological Control, Progress in Biological Control*, vol. 12; Mérillon, J. M., Ramawat, K. G., Eds., 2012, 139–160.

Rembold, H. Secondary Plant Compounds in Insect Control with Special Reference to Azadirachtin. *Adv. Invertebrate Reprod.* **1994**, *3*, 481–491.

Rice, E. L. *Allelopathy*, 2nd ed. Academic Press: New York, 1987.

Ritsuo, N. Chemical Ecology of Insect–Plant Interactions: Ecological Significance of Plant Secondary Metabolites. *Biosci. Biotechnol. Biochem.* **2014**, *78*(1), 1–13.

Rodriguez, E. Insect Feeding Deterrents from Semi-arid and Arid Land Plants. In *Bioregulators for Pest Control*; Hedin, P. A. Ed.; ACS Symposium Series 276, American Chemical Society: Washington, DC, 1985; pp 447–453.

Rossiter, M.; Gershenzon, J.; Mabry, T. J. Behavioural and Growth Responses of Specialist Herbivores, Homoeosoma Electelum to Major Terpenoid of its Host *Helianthus* spp. *J. Chem. Ecol.* **1986**, *12*, 1505–1521.

Salvadores, U. Y.; Silva, A. G.; Tapia V. M.; Hepp, G. R. Polvos de especias aromáticas para el control del gorgojo del maíz, *Sitophilus zeamais* Motschulsky, en trigo almacenado. *Agric. Téc. (Chile)* **2007**, *67*, 147–154.

Saito, K. Sulfur Assimilatory Metabolism. The Long and Smelling Road. *Plant Physiol.* **2004**, *136*, 2443–2450.

Singh Rattan, R.; Anuradha, S. Plant Secondary Metabolites in the Sustainable Diamondback Moth (*Plutella xylostella* L.) Management. *Indian J. Fundam. Appl. Life Sci.* **2011**, *1*(3), 295–309.

Silva, G.; Lagunes, A. Y.; Rodríguez, J. Control de *Sitophilus zeamais* (Coleoptera: Curculionidae) con Polvos Vegetales Solos y en Mezclas con Carbonato de Calcio en Maíz Almacenado. *Cienc. Investig. Agr.* **2003**, *30*, 153–160.

Sreevidya, V. S.; Srinivasa, R. C.; Rao, C.; Sullia, S. B.; Ladha, J. K.; Reddy, P. M. Metabolic Engineering of Rice with Soyabean Isoflavone Synthase for Promoting Nodulation Gene Expression in Rhizobia. *J. Exp. Bot.* **2006**, *57*(9), 1957–1969.

Taiz, L.; Zeiger, E. *Plant Physiology*, 4th ed. Sinauer Associates Inc. Publishers: Massachusetts, 2006.

Talalay, P.; Fahey, J. W. Phytochemicals from Cruciferous Plants Protect Against Cancer by Modulating Carcinogenic Metabolism. *J. Nutr.* **2001**, *131*, 3027–3033.

Thebtaranonth, C.; Thebtaranonth, Y.; Wanauppathamkul, S.; Yuthavong, Y. Antimalarial Sesquiterpenes from Tubers of *Cyperus rotundus*: Structure of 10,12-Peroxycalamenene, a Sesquiterpene Endoperoxide. *Phytochemistry* **1995**, *40*, 125–128.

Thomma, B. P. H. J.; Cammue, B. P. A.; Thevissen, K. Plant Defenses. *Planta* **2002**, *216*(2), 193–202.

Turlings, T. C. J.; Loughrin, J. H.; Mccall, P. J.; Roese, U. S. R.; Lewis, W. J; Tumlinson, J. H. How Caterpillar-Damaged Plants Protect Themselves by Attracting Parasitic Wasps. *Proc. Natl. Acad. Sci. U.S.A.* **1995**, *92*, 4169–4174.

Van Loon, L. C.; Pierpoint, W. S.; Boller, T.; Conejero, V. Recommendations for Naming Plant Pathogenesis-Related Proteins. *Plant Mol. Biol. Rep.* **1994**, *12*, 245–264.

Van Etten, H. D.; Mansfield, J. W.; Bailey, J. A.; Farmer, E. E. Two Classes of Plant Antibiotics: Phytoalexins Versus "Phytoanticipins". *Plant Cell* **1994**, *6*, 1191–1192.

Van Steveninck, R. F. M.; Van Steveninck, M. E. Abscisic Acid and Membrane Transport. In *Abscisic Acid*; Addicott, F. T. Ed.; Praeger: New York, 1983; pp 171–235.

Vickery, B.; Vickery, M. L. *Secondary Plant Metabolism*. University Park Press: Baltimore, 1981.

Viglianco, A. I.; Novo, R.; Cragnolini, C. Y.; Nassetta, M. Actividad Biologica de Extractos crudos de *Larrea divaricata* Cav. y *Capparis atamisquea* Kuntze sobre *Sitophilus oryzae* (L.). *Agriscientia* **2006**, *23*, 83–89.

Wakabayashi, N.; Wu, W. J.; Waters, R. M. Celangulin: A Non-Alkaloidal Insect Antifeedant from Chinese Bittersweet *Celasatrus angulatus*. *J. Nat. Prod.* **1988**, *51*, 537–542.

Wallsgrove, R.; Benett, R.; Kiddle, G.; Bartlet, E.; Ludwig-Mueller, J. Glucosinolate Biosynthesis and Pest Disease Interactions. Proceedings of the 10th International Rapeseed Congress, Canberra, Australia. 1999.

Wanjala, W.; Cornelius, T. A.; Obiero, G. O.; Lutta, K. P. Antifeedant Activities of the Erythrinaline Alkaloids from *Erythrina latissima* against *Spodoptera littoralis* (Lepidoptera noctuidae) *Rec. Nat. Prod.* **2009**, *32*, 96–10.

Wheeler, D.; Isman, M. Antifeedant and Toxic Activity of *Trichilia americana* Extract against the Larvae of *Spodoptera litura*. *Entomol. Exp. Appl.* **2001**, *98*, 9–16.

Wuyts N. De waele D. Swennen R. Extraction and Partial Characterization of Polyphenol Oxidase from Banana (*Musa acuminate* Grandr Naine) Roots. *Plant Physiol. Biochem.* **2006**, *44*, 308–314.

Wretblad, S.; Dixelius, C. B-genome Derived Resistance to *Leptosphaeria maculans* in Near Isogenic *Brassica napus* Lines Is Independent of Glucosinolate Profile. *Physiol. Plant.* **2000**, *110*, 461–468.

Zukalova, H.; Vasak, J. The Role and Effects of Glucosinolates of *Brassica* species: *A Review. Rostlinna Vyroba* **2002**, *48*(4), 175–180.

CHAPTER 6

SECONDARY METABOLITES IN PATHOGEN-INDUCED PLANT DEFENSE

MOHAMMAD ANSAR[1], ABHIJEET GHATAK[1*], LAJJA VATI GHATAK[2], ASHWATHNARAYAN SRINIVASARAGHAVAN[1], REKHA BALODI[3], and CHANDRAMANI RAJ[4]

[1]Plant Pathology, BAC, Bihar Agricultural University, Sabour, Bhagalpur 813210, Bihar, India
[2]Plant Breeding and Genetics, BAC, Bihar Agricultural University, Sabour, Bhagalpur 813210, Bihar, India
[3]Plant Pathology, COA, Govind Ballabh Pant University of Agriculture and Technology, Udham Singh Nagar, Pantnagar 263145, Uttarakhand, India
[4]Dryland Cereal Pathology, International Crop Research Institute for Semi-Arid Tropics, Patancheru, Hyderabad 502324, Telangana, India
*Corresponding author, E-mail: ghatak11@gmail.com.

CONTENTS

ABSTRACT

The attack of pathogens on the plants also stimulates the release of secondary metabolites. In fruit and vegetables, there is a wide variety of pathogens that pose quality threat on nutrition, organoleptic degradation, and shortened shelf life. In response to foreign hurdle, the microorganisms have developed a system for their defense against various unwanted elements. The moment pathogen approaches to host surface, the plant expresses characters of several naturally occurring structural and biochemical barriers resisting penetration. This chapter highlights the release of biochemicals on being attacked by pest.

6.1   INTRODUCTION

The phytopathogen is one of the major biotic stresses that contribute substantially to overall loss among agricultural crops (Agrios, 2005). According to a conservative estimate, almost 10–20% of staple foods and cash crops are destroyed by phytopathogenic fungi (Hewitt, 2000). These defense systems have been formed depending upon the basic structural construction of the organism and their need (Barth et al., 2009).

Although, analysis of most of the host–pathogen relationships exhibited the pattern of pathogenesis for a pathosystem; the plants change their constructs during the pathogen infection and exhibit defense mechanism either by modification in cellular structure or by secreting a special chemical (Mazid et al., 2011). The structural defense includes tylosis in vascular wilt, formation of cork layer in European plum, abscission layer in peach, and so on. Similarly the biochemical secretions involve phytoalexin (ipomeamarone) in sweet potato, saponin (tomatin) in tomato; likewise phenolic compound (tannin) and fatty acid (dienes) in young fruits prevent infection to *Botrytis cinerea*. Both the structural and biochemical barriers are developed before and after infection. Preformed chemical substances such as phenolic and polyphenol compounds are always present in the plant system and therefore play a crucial role in nonhost resistance to plant pathogenic organisms. Production of large number of organic compounds (which are of various types) is found in plants.

Here in this chapter, we are going to focus on the biochemical defense highlighting postinfection secondary metabolites. The secondary metabolites often have critical role in survivability, and fecundity (Hadacek, 2002). Secondary metabolites neither directly interfere the usual growth of the organism nor do they interrupt the reproduction process. However, lack of primary metabolites may result in sudden death of the organism. In short, primary metabolites are the cause of intrinsic (physiological) function whereas the secondary metabolites are responsible for ecological function. Moreover, a definite primary metabolite is found in many organisms; however, a secondary metabolite is generally taxonomically bounded by nature that means they are confined to bacteria, fungi, plants, and others. Within phylogenetic group, a narrow set of species is governed by a particular secondary metabolite. They are low molecular weight compound that are chemically organic.

The excessive use of pesticides in order to reduce crop losses has several environmental and health issues. At this juncture, the secondary metabolites are the new hope for plant disease management in near future. Here, we present a brief review of the work done on secondary metabolites, their role in plant disease resistance, and the ways in which it can be exploited in plant disease management.

6.2 WHAT ARE SECONDARY METABOLITES?

Plants, particularly agricultural crops, are usually infected by several phytopathogens. In response to this pathogenic process, the plants synthesize special chemical(s), which saves the plant from several phytopathogens like viruses, phytoplasmas, bacteria, fungi, and others. This synthesized chemical contains numerous varieties of organic molecules, called secondary metabolites. The secondary metabolites are low molecular weight compound. Such chemicals are often found as end-product of nitrogen metabolism. As a result of various types of interactions frequently operated between plants and other organisms, the plant-generated secondary metabolites have a wide range of biological performance. In terms of a plant system, the synthesized compound that has potential role to interact the plant with its environment and does not interfere with normal life processes of the plant is known as secondary metabolite. Such chemicals are used as weapon by the plant to fight against a number of pests (Dixon, 2001). Thus, production of secondary metabolites leads to defense in plants.

6.3 GENERAL SIGNIFICANCE OF SECONDARY METABOLITES

The science of secondary metabolites previously unveiled to the modern science. As a common belief, these compounds were thought to be the end-product(s) of metabolism, which is often function less. The upgradation in our knowledge explains various biological functions of these materials in plants. The function of the secondary metabolites is in both ways: attractive and protective. These chemicals attract the pollinators and seed-dispersal animals for dissemination of a plant species. They attract the disseminating agents by means of color, odor, and taste of the chemical. Protection is performed by the antimicrobial (pathogen) and antifeeding (herbivores) activities (Picman, 1986). Moreover, it acts as precursor to the physical defense system, and its toxicity is associated with the retardation of the growth of pathogens. The purpose of secondary metabolites could be understood as an agent of plant-to-plant competition and plant-to-microbe symbiosis because of their ecological function (Dixon, 2001; Harborne, 2001). Therefore, secondary metabolites play the guiding role in order to compete and survive the plants to its ecology. Impact of secondary metabolites on a plant species is, thus, a deciding factor for its existence. Theme of secondary metabolites is untouched to agriculture system, particularly to postharvest management. However, its role is identified for different pathosystems under disease management (Fig. 6.1). Moreover, the importance of secondary metabolites in pathogenesis is well understood. During pathogenesis, these chemicals are of either microbial origin or plant origin. Secondary metabolites as product of microbe are used to manage other pathogen. For example, lytic peptides from different plant growth-promoting rhizobacteria are effective against various plant pathogens. The plant-originated secondary metabolites are also utilized to protect crops against several phytopathogens. In few cases, some of the important crop plants are artificially selected to produce relatively low levels of secondary metabolites. Engineering of these metabolites could be utilized for disease management (Verpoorte & Memelink, 2002). On the view of plant pathology, secondary metabolites bear antimicrobial property. According to performance, they may be divided into two broad categories, that is, pathogenesis and management of disease (Fig. 6.1).

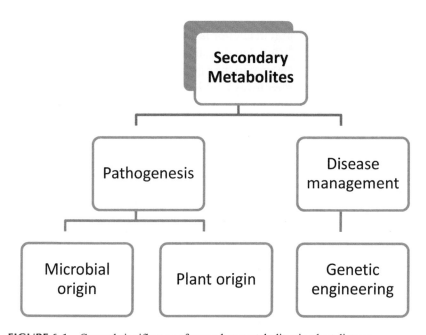

FIGURE 6.1   General significance of secondary metabolites in plant disease.

6.4   PLANT–MICROBE INTERACTION

The phytopathogens cause numerous diseases by interfering general physiology of the plant. At initial phase, it regulates the defense responses in plant including a wide range of molecules present in little amount in the system. This range of molecules is known as elicitor that is produced by the phytopathogen and triggers defense response in plant (Agrios, 2005). They may be host-specific: if the elicitor is produced by the *avr* (avirulence) gene of pathogen and interacts with R (resistance) gene of plant. However, some of the elicitors are general that are recognized by a few number of plants.

Elicitors are chemicals or biofactors from various sources that can induce physiological changes of the target living organism (Fig. 6.2). Therefore, the pathogen-produced elicitors for a plant include chemicals from different sources that can trigger physiological and morphological responses, and phytoalexin (as discussed above) accumulation. It includes (1) nonliving elicitors such as metal ions and inorganic compounds, and

(2) living elicitors from fungi, bacteria, viruses, or herbivores, and so on. One of the examples of protein elicitor is a bacterial flagellin (Bauer et al., 2001). Elicitors or avirulence determinants must be recognized by plant receptors or R proteins localized to the plasma membrane or the cytoplasm before initiating signaling pathways, which lead to defense reactions such as synthesis of PR (pathogenesis-related) proteins, or defense secondary metabolites. Molecular recognition and physical interaction between elicitor signal molecules and specific plant receptors are complex processes but are required for specific elicitor signal transduction (Qing et al., 2007).

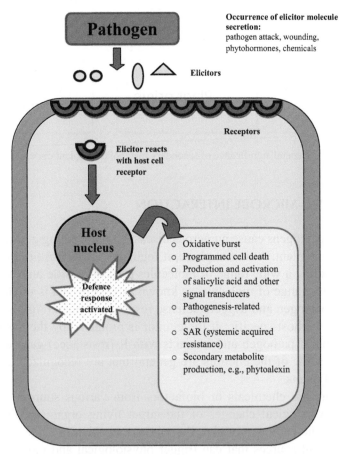

FIGURE 6.2 Schematic representation of activation of defense response in plants upon pathogen onset.

Phytopathogen producing a type-elicitor may or may not be recognized by the host plant (Gurr & Rushton, 2005). Recognition of elicitor is operated by the receptors aligned in plasma membrane of the plant cell. A definite receptor is made to identify a particular type-elicitor. Figure 6.2 captures the process of elicitor recognition and initiation of defense. If the plant receptor does not recognize the type-elicitor produced by phytopathogen, infection will comfortably happen. But when the plant receptor recognizes an elicitor, there is formation of elicitor–receptor complex, which sends information to the host nucleus. Thus, the defense response is activated and defense mechanism starts. Upon onset of pathogen on the host surface, the pathogen releases many elicitor molecules. These molecules are of different types and could be generated by wounding, application of phytohormones or other chemicals beside the pathogen attachment to the surface. The elicitors are recognized by the receptors aligned in the plasma membrane of the host plant. The plant defense after recognition of elicitor is reacted with oxidative burst, programmed cell death, activation of systemic acquired resistance, and others; and thus, the plant becomes protected.

6.5 CATEGORY OF SECONDARY METABOLITES

Various secondary metabolites (number may exceed to thousand) are produced in plants and accumulated in very low quantity (Croteau et al., 2000; Dewick, 2002). They are essential for survival and generation of offspring of a producing plant but not required for routine growth and other activities (Hadacek, 2002). Secondary metabolites synthesized in plants follow specific pathway. Different secondary metabolites are produced on different sites in a plant, and even, diverse compounds are synthesized in different plant species. However, some of the chemicals are synthesized in all kinds of tissues in a plant while rest are tissue-specific or sometimes cell-specific or organ-specific (Yazdani et al., 2011). The plant-synthesized hydrophilic secondary metabolites are mostly accumulated in vacuole while the lipophilic secondary metabolites are commonly sequestered in resin ducts, laticifers, oil cells, trichomes, or in the cuticle over epidermis (Engelmeier & Hadacek, 2006). Different categories of secondary metabolites are generated in plant, which is briefly discussed in Figure 6.3. Three types of plant-produced secondary metabolites are secreted in response to defense against pathogenic infection. They are categorized under (1) phenolics, (2) terpenoids, and (3) nitrogen-and sulfur-containing compounds, which are briefly discussed below.

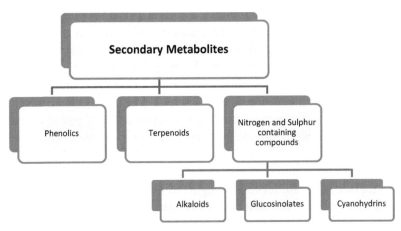

FIGURE 6.3   A broad category of secondary metabolites.

6.5.1   PHENOLICS AND ANTIMICROBIAL ACTIVITY

Shikimic acid pathway leads to synthesis of phenolics in plants which play significant role in plant defense mechanism. Phenolics determine pigmentation, growth, and reproduction, and resistance to phytopathogens. It is therefore well understood that they participate in the evolutionary processes in order to better adaptation of characters. Chemically, phenols are soluble in polar-organic solvents unless they are completely esterified, etherified or glycosylated. Plant phenols are of two types: (1) preformed—present prior to infection and (2) induced—synthesized in response to elicitors (Fig 6.2).

Induced phenolics (e.g., phytoalexins) are synthesized after recognition of elicitors from pathogen onset, application of hormones or chemicals, injury, and others. Examples of different phenols are provided in Table 6.1. Polyphenols play a role in the plant defense against fungal and other microbial pathogens. Likewise, flavonoids regulate pigmentation and thus defense of the plant. Some of the isoflavonoids help to increase nitrogen-fixing nodules by root-associated rhizobacteria. Chlorogenic acid, important phenolics of potato, is highly effective against filamentous bacterium

Streptomyces scabies (Table 6.2). In grapevine, a phytoalexin, trans-resveratrol is produced against the postharvest fungal pathogens such as *B. cinerea*, and similarly, delta-viniferin—another grapevine phytoalexin produced after downy mildew fungal infection by *Plasmopara viticola*. Tanic acid is synthesized during viral infection in tobacco. The induced phenols may synthesize constitutively; however, an increased synthesis could be recognized for plants challenged by pathogens.

TABLE 6.1 Major Secondary Metabolites and Their Function.

Class	Examples	Major function
Terpenoids		
Monoterpene	Pinene, citronellol, limonene	Plant–plant interaction
Sesquiterpene	Abscisic acid	Plant growth hormone
	Ryanodine	Insecticides
Diterpene	Giberelic acid	Plant growth hormone
Triterpene	Azadirachtin	Antifeedant, disruption of oviposition, reduction in fecundity
	Saponins, glycosidic tri-terpenoids	Antifungal
Phenolics		
Simple phenols	Catechol	Antifungal
Stilbenes	Resveratrol	Antifungal
Flavonoids	Quercetin	Pigmentation and defense
Polyphenols	Lignins	Blocking of pathogenic growth, disease resistance, feeding repellents
	Condensed tannins	
Furanocoumarins	Umbelliferone	Fungal defense
Isoflavonoids	Naringenin	Promoting the formation of nitrogen-fixing nodules by symbiotic rhizobia
Nitrogen-and sulfur-containing compounds		
Alkaloids	Pyrrolizidine	Antimicrobial
Cyanogenic glucosides	Amygdalin	Broad-spectrum defense
Phytochelatins	Glutathione	Cellular antioxidant
Glucosinolates	Allyl-*cys*-sulfoxides	Antifungal

TABLE 6.2   List of Different Secondary Metabolites and Their Role in Disease Resistance.

Host	Chemical	Effective against pathogen
Terpenoids		
Tomato	Tomatine	Antifungal
Oat	Avenacin	Antifungal
Phenolics		
Potato	Chlorogenic acid	*Streptomyces scabies*
Seedling of potato	Chlorogenic acid	*Verticillium wilt*
Grapevine	Trans-resveratrol	*Botrytis cinerea*
	Delta-viniferin	*Plasmopara viticola*
Apple	Phloridzin	*Venturia inequalis*
Tobacco	Tannic acid	*Tobacco Mosaic Virus*
Avocado	Manoens and dienes	*Colletorichum gleosporides*
Papaya	Danielone	*Colletorichum gleosporides*
Coloured onion bulb	Procatechuic acid and catechol	*Colletotrichum cicinans*
Nitrogen-and sulfur-containing compounds		
Sorghum	Dhurrin	Fungi and nematodes
Rapeseed	Glucosinolates	*Sclerotinia sclerotiorum*
Wheat	Thionins	*Fusarium culmorum*

6.5.2 TERPENOIDS AND ANTIMICROBIAL ACTIVITY

Terpenoids are low molecular weight compounds which are volatile in nature and composed of 5-C isopentanoid units (Gershenzon & Croteau, 1991; Grayson, 1998). They constitute the largest class of secondary metabolites having properties, namely, feeding deterrents against herbivores and insects, ovipositional inhibitors of insects and sometimes antifungal also (Gershenzon & Croteau, 1991). In plant defenses, terpenoids have been an considered to be important but their active role in resistance against pathogenic reaction is still not well known. Terpenoids (mono-, di-, tri- and sesqui-terpenes) are normally the most hydrophobic substances and accumulate in resin ducts, oil cells, or glandular trichomes (Wink, 2010). The mono-terpenes carvacrol and thymol are the main antifungal compounds found in *Satureja thymbra* and *Thymbra spicata*, respectively (Müller-Riebau et al., 1997). The investigations on these secondary metabolites clearly showed effective antifungal activity when evaluated with some commercial fungicides. The strong antifungal activity of carvacrol and thymol was also confirmed on the ubiquitous phytopathogenic fungi *B. cinerea* (Bouchra et al., 2003; Camele et al., 2012; Tsao & Zhou, 2000). The *in-vitro* gaseous state mono-terpene γ-terpinene assessment revealed a high inhibitory activity against *B. cinerea* (Espinosa-Garcia & Langenheim, 1991). Additionally, the mono-terpene citral has a potent antimicrobial compound against *Penicillium italicum* (Saddiq & Khayyat, 2010), *B. cinerea* (Tsao & Zhou, 2000), and *Colletotrichum acutatum* (Alzate et al., 2009). Some of terpenes play a vital role as plant growth hormones. Saponin, a tri-terpene, has high antifungal property, which is synthesized before infection (Table 6.1). The well-known examples of saponins are tomatine in tomato and avenacin in oat (Table 6.2).

6.5.3 NITROGEN-AND SULFUR-CONTAINING COMPOUNDS AND ANTIMICROBIAL ACTIVITY

The compounds containing nitrogen and sulfur fall in the third important group of secondary metabolites. Nitrogen-and sulfur-containing secondary metabolites are further grouped into alkaloids, glucosinolates and cyanohydrins (Fig. 6.3). The member of these groups is principally synthesized from amino acids (Rosenthal, 1992; Van Etten et al., 2001). The experiments under controlled condition demonstrate change in expression of secondary metabolites in order to confirm their role in defense (Mansfield, 2000; Mes et al., 2000).

Nitrogen-containing antimicrobial compounds have wide range of functions (Table 6.1). Poulton (1990) found that dhurrin, a nitrogenous alkaloid, present in sorghum is highly effective against soilborne pathogens (Table 6.2). Glutathione, glucosinolates, phytoalexins, thionins, defensins, and allinin are the major sulfur-containing compounds involved in plant defense mechanism against phytopathogens (Grubb & Abel, 2006; Halkier & Gershenzon, 2006). Most of them are governed by the systemic acquired resistance (Bloem et al., 2005). Glutathione, an example of phytochelatin, has additional role of cellular antioxidant in plant (Table 6.1). Glucosinolates and thionins can control *Sclerotinia sclerotiorum* and *Fusarium culmorum* for rapeseed and wheat, respectively (Table 6.2).

6.6 PHYTOALEXIN: PLANT DEFENSE WEAPON POST INFECTION

Phytoalexin, a well-identified molecule having antimicrobial property, is synthesized only at post-infection condition and gets accumulated around both susceptible and resistant necrotic spots (Agrios, 2005). However, phytoalexin production has not been detected during compatible biotrophic interactions. Mostly, they are common both in gymnosperms and angiosperms. Phytoalexins belong to three major groups, that is, terpenoids, phenylpropanoids, and fatty acid derivatives. Rice phytoalexins, momilactones A and B, accumulated in maximum concentration at 72–96-h post infection by the rice blast fungus *Magnaporthe oryzae*. Yamada et al. (1993) showed that inhibition of growth in rice blast fungus could be obtained at very low concentrations of momilactones. These chemicals require *de-novo* expression of the involved enzymes in their biosynthetic pathway. Moreover, application of silicon enhances accumulation of phytoalexin that saves rice from blast disease (Rodrigues et al., 2004). A large number of crops induce phytoalexin upon interaction with its pathogen. The list of phytoalexins for several pathogenic interactions is presented in Table 6.3. Within a family, several crops may be infected and thus induction of phytoalexin takes place. A phytoalexin (e.g., capsidol) may be effective against different pathosystems, for example, fungus and

TABLE 6.3 Phytoalexins Produced During Host–Pathogen Interaction.

Family of crop	Crop	Pathogen	Phytoalexin
Solanaceae	Pepper	*Colletotrichum capsici*	Capsidol
	Tobacco	Tobacco necrosis virus	Capsidol
	Tobacco	Tobacco mosaic virus	Glutinosone
	Potato	*Phytophthora infestans*	Rishitin
Convolvulaceae	Sweet potato	*Ceratocystis fimbriata*	Ipomeamarone
Malvaceae	Cotton	*Verticillium albo-atrum*	Gossypol
Leguminosae	Pea	*Monilinia fruiticola*	Pisatin
	Pea	*Botrytis fabae*	Wyerone
	Chickpea	*Ascochyta rabiei*	Cicerin
	Pea, chickpea	*Fusarium* sp.	Maackiain
	French bean	*Sclerotinia fructigena*	Phaseolin
Orchidaceae	Orchid	*Mycorrizal fungi*	Orchinol and hircinol

virus. The phytoalexins may also be induced for the beneficial interaction. Within a host, various phytoalexins are synthesized for different pathosystems (e.g., pea). In orchid, orchinol and hircinol are synthesized against mycorrhizal fungi.

6.7 CONCLUSION

Pathogenic microorganisms constitute one of the main infectious agents, inciting deformities during different stages of plant growth. Commonly, phytopathogenic elements are managed by various systemic agrochemicals; however, the use of these is progressively restricted due to both the resistance development and effect on the environment and human health. Therefore, there is an urgent need for novel natural pesticides. Interestingly, plants are a loaded source of bioactive secondary metabolites of wide variety such as terpenoids, saponins, alkaloids, and other compounds. These are found to inhibit the growth of phytopathogens under *in-vitro* condition. Consequently, secondary metabolites with an antifungal activity represent a substitute for achieving a sustainable control of phytopathogens and to reduce the heavy application of systemic chemicals. Plant antifungal metabolites may be preformed inhibitors that are present constitutively in healthy plants (phytoanticipins, *not discussed*) or they may be synthesized *de-novo* in response to pathogen attack or another stress conditions (phytoalexins), which could be considered as a precursor for developing better fungicidal molecules.

Ever-increasing global human population has mounted an enormous pressure on agro-ecosystem. There is a need to enhance the production with limited land resources. In this context, preventing the losses caused by the plant diseases is need of the hour. To save the loss, in general, the growers are depending more and more on application of pesticides, which is raising severe environmental concerns. Exploitation of such disease management options, which have less or no adverse impact on environment are very much needed. Secondary metabolites produced in plants are new hope in this direction. Secondary metabolites, although not essential, are very important in enhancing the survival ability of the plant. There is an array of secondary metabolites with diverse mode of actions involved in plant defense mechanism leading to disease resistance. The phenolic compounds are the plant defense weapons produced in order to guard the pathogen attack, in contrary, phytoalexins mount the defense against the attacking pathogen. This basic inherent strategy present in plants provides a huge opportunity to exploit these secondary metabolites to enhance the disease resistance through genetic engineering. Production of secondary metabolites is performed artificially with biotechnological inputs. Plant cell culture is found as an alternative for this purpose, but till date, it has very limited success particularly for commercial use. The reason is the lack of knowledge that must flow through the graduation level. In this chapter, we tried to present a selection of antifungal substances induced in plants during fungal colonization that can be potentially be used for fungal disease management in various crops.

KEYWORDS

- elicitor
- phytoalexins
- phytopathogen
- plant–microbe interaction
- secondary metabolites

REFERENCES

Agrios, G. N. *Plant Pathology*. Academic Press: San Diego, 2005.

Alzate, O.; Diego, A.; Mier, M.; Gonzalo, I. L.; Afanador, K.; Durango, R.; Diego, L.; Garcia, P.; Carlos, M. Evaluation of Phytotoxicity and Antifungal Activity against *Colletotrichum acutatum* of Essential Oils of Thyme (*Thymus vulgaris*), Lemongrass (*Cymbopogon citratus*), and Their Main Constituents. *Vitae* **2009**, *16*, 116–125.

Barth, M.; Thomas, R.; Hankinson, Zhuang, H.; Breidt, F. Microbiological Spoilage of Fruits and Vegetables. In *Compendium of the Microbiological Spoilage, Food Microbiology and Food Safety*; Sperber, W. H., Doyle, M. P., Eds.; 2009, pp 135–183.

Bauer, Z.; Gómez-Gómez, L.; Boller, T.; Felix, G. Sensitivity of Different Ecotypes and Mutants of *Arabidopsis thaliana* toward the Bacterial Elicitor Flagellin Correlates with the Presence of Receptor-Binding Sites. *J. Biol. Chem.* **2001**, *276*(49), 45669–45676.

Bloem, E.; Haneklaus, S.; Schnug, E. Significance of Sulphur Compounds in the Protection of Plants against Pests and Diseases. *J. Plant Nutr.* **2005**, *28*, 763–784.

Bouchra, C.; Achouri, M.; Idrissi Hassani, L. M.; Hmamouchi, M. Chemical Composition and Antifungal Activity of Essential Oils of Seven Moroccan Labiatae against *Botrytis cinerea* Pers: Fr. *J. Ethnopharmacol.* **2003**, *89*, 165–169.

Camele, I.; Altieri, L.; De Martino, L.; De Feo, V.; Mancini, E.; Rana, G. L. In Vitro Control of Post-Harvest Fruit Rot Fungi by Some Plant Essential Oil Components. *Int. J. Mole. Sci.* **2012**, *13*, 2290–2300.

Croteau, R.; Kutchan, T.; Lewis, N. Natural Products (Secondary Metabolites). In *Biochemistry and Molecular Biology of Plants*; Buchanan, B., Gruissem, W., Joneas, R. Eds.; American Society of Plant Biologists: Rockville, MD, 2000; pp 1250–1268.

Dewick, P. M. *Medicinal Natural Products: A Biosynthetic Approach*, 2nd ed.; John Wiley and Sons Ltd.: Chichester, 2002, UK; pp 507.

Dixon, R. A. Natural Products and Plant Disease Resistance. *Nature* **2001**, *411*, 843–847.

Engelmeier, D.; Hadacek, F. Antifungal Natural Products: Assays and Applications. In *Naturally Occurring Bioactive Compounds*; Rai, M., et al., Eds.; Elsevier Science Ltd. 2006, pp 423–467.

Espinosa-Garcia, F. J.; Langenheim, J. H. Effects of Sabinene and γ-Terpinene from Coastal Redwood Leaves Acting Singly or in Mixtures on the Growth of Some of Their Fungus Endophytes. *Biochem. Syst. Ecol.* **1991**, *19*, 643–650.

Gershenzon, J.; Croteau, R. Terpenoids. In *Herbivores Their Interaction with Secondary Plant Metabolites, Vol I: The Chemical Participants*, 2nd ed.; Rosenthal, G. A., Berenbaum, M. R. Eds.; Academic Press: San Diego, 1991, pp 165–219.

Grayson, D. H. Monoterpenoids. Natural Product Reports, 1998, 5, 497–521.

Grubb, C.; Abel, S. Glucosinolate Metabolism and Its Control. *Trends Plant Sci.* **2006**, *11*, 89–100.

Gurr, S. J.; Rushton, P. J. Engineering Plants with Increased Disease Resistance: How Are We Going to Express It? *Trends Biotechnol.* **2005**, *23*(6), 283–290.

Hadacek, F. Secondary Metabolites as Plant Traits: Current Assessment and Future Perspectives. *Crit. Rev. Plant Sci.* **2002**, *21*, 273–322.

Halkier, B. A.; Gershenzon, J. Biology and Biochemistry of Glucosinolates. *Ann. Rev. Plant Biol.* **2006**, *57*, 303–333.

Harborne, J. B. Twenty-Five Years of Chemical Ecology. *Nat. Prod. Rep.* **2001**, *18*, 361–379.

Hewitt, G. New Modes of Action of Fungicides. *Pesticide Outlook* **2000**, *11*, 28–32.

Mansfield, J. W. Antimicrobial Compounds and Resistance. The Role of Phytoalexins and Phytoanticipins. In *Mechanisms of Resistance to Plant Diseases*; Slusarenko, A., Fraser, R., Van Loon, L., Eds.; Kluwer Academic Publishers: Netherlands, 2000; pp 325–370.

Mazid, M.; Khan, T. A.; Mohammad, F. Role of Secondary Metabolites in Defense Mechanisms of Plants. *Biol. Med.* **2011**, *3*(2), 232–249.

Mes, J. J.; Van Doorn, A. A.; Wijbrandi, J.; Simons, G.; Cornelissen, B. J. C.; Haring, M. A. Expression of the *Fusarium* Resistance Gene I-2 Colocalizes with the Site of Fungal Containment. *Plant J.* **2000**, *23*, 183–193.

Müller-Riebau, F. J.; Berger, M.; Yegen, O.; Cakir, C. Seasonal Variations in the Chemical Compositions of Essential Oils of Selected Aromatic Plants Growing Wild in Turkey. *J. Agric. Food Chem.* **1997**, *45*, 4821–4825.

Picman, A. K. Biological Activities of Sesquiterpene Lactones. *Biochem. Syst. Ecol.* **1986**, *14*, 255–281.

Poulton, J. E. Cyanogenesis in Plants. *Plant Physiol.* **1990**, *94*, 401–405.

Qing, Y.; Xiu-fen, Y.; Ying, L.; Feng, X.; Zheng, L.; Jing-jing, Y.; De-wen, Q. Expression of A *Magnaporthe grisea* Elicitor and Its Biological Function in Activating Resistance in Rice. *Rice Sci.* **2007**, *14*(2), 149–156.

Rodrigues, F. Á.; McNally, D. J.; Datnoff, L. E.; Jones, J. B.; Labbé, C.; Benhamou, N.; Menzies, J. G.; Bélanger, R. R. Silicon Enhances the Accumulation of Diterpenoid Phytoalexins In Rice: A Potential Mechanism for Blast Resistance. *Phytopathology* **2004**, *94*, 177–183.

Rosenthal, G. A. Purification and Characterization of the Higher Plant Enzyme L-Canaline Reductase. *Proc. Natl. Acad. Sci. U.S.A.* **1992**, *89*, 1780–1784.

Saddiq, A. A.; Khayyat, S. A. Chemical and Antimicrobial Studies of Monoterpene: Citral. *Pestic. Biochem. Phys.* **2010**, *98*, 89–93.

Tsao, R.; Zhou, T. Antifungal Activity of Mono-Terpenoids against Postharvest Pathogens *Botrytis cinerea* and *Monilinia fructicola*. *J. Essen. Oil Res.* **2000**, *12*, 113–121.

Van Etten, H. D.; Mansfield, J. W.; Bailey, J. A.; Farmer, E. E. Two Classes of Plant Antibiotics: Phytoalexins versus "Phytoanticipins". *Plant Cell* **1994**, *6*, 1191–1192.

Van Etten, H.; Temporini, E.; Wasmann, C. Phytoalexin (and Phytoanticipin) Tolerance as A Virulence Trait: Why Is It Not Required by All Pathogens? *Physiol. Mol. Plant Pathol.* **2001**, *59*, 83–93.

Verpoorte, R.; Memelink, J. Engineering Secondary Metabolite Production in Plants. *Curr. Opin. Biotechnol.* **2002**, *13*, 181–187.

Wink, M. Functions and Biotechnology of Plant Secondary Metabolites. *Annu. Plant Rev.* **2010**, *39*.

Yamada, A.; Shibuya, N.; Kodama, O.; Akatsuka, T. Induction of Phytoalexin Formation in Suspension-Cultured Rice Cells by *N*-Acetyl-chito-oligosaccharides. *Biosci. Biotech. Biochem.* **1993**, *57*, 405–409.

Yazdani, D.; Tan, Y. H.; Zainal Abidin, M. A.; Jaganath, I. B. A Review on Bioactive Compounds Isolated from Plants against Plant Pathogenic Fungi. *J. Med. Plants Res.* **2011**, *5*(30), 6584–6589.

CHAPTER 7

POLYHYDROXYALKANOATES: A VALUABLE SECONDARY METABOLITE PRODUCED IN MICROORGANISMS AND PLANTS

IDA IDAYU MUHAMAD[1,2*], FARZANEH SABBAGH[1], and NORSUHADA ABDUL KARIM[1]

[1]*Bioprocess Engineering Department, Faculty of Chemical Engineering, Universiti Teknologi Malaysia, 81310 Johor Bahru, Johor, Malaysia*

[2]*IJN-UTM Cardio Engineering Centre, V01 FBME, Universiti Teknologi Malaysia, 81310 Johor Bahru, Johor, Malaysia*

*Corresponding author, E-mail: idayu@cheme.utm.my

CONTENTS

ABSTRACT

PHAs (polyhydroxyalkanoates) are the polymers of hydroxyalkanoates and these polymers agglomerate as carbon and energy or to reduce storage materials in the bacteria. It has a promising importance in the various applications such as agricultural (e.g., controlled release of insecticides, mulch films), environmental (binders, biocomposites, adhesives), food packaging, flexible packaging, and biomedical applications (medical devices, scaffolds for tissue engineering applications). This chapter discusses the importance of PHAs.

7.1   INTRODUCTION: AN OVERVIEW

Secondary metabolites are an extremely diverse and important class of natural products. Polyhydroxyalkanoates (PHAs) are one of the biopolymers which are a class of secondary metabolites produced in the microorganism or plants. PHAs are stored in the bacterial cytoplasm as inclusion bodies and they are synthesized and accumulated intracellular as clear granules ((Koutinas et al., 2014; Salehizadeh & Van Loosdrecht, 2004). PHAs family can be classified according to the chain length of the branching polymers. Short-chain-length PHAs (scl-PHAs) are composed of 3–5 carbon atoms, while medium-chain-length- and long-chain-length PHAs (lcl-PHAs) consist of 6–14 and over 14 carbon atoms, respectively. The most extensively studied member of the PHA family is poly(3-hydroxybutyrate) (PHB). It was in the mid-1920s, Lemoigne at the Pasteur Institute in Paris, first, identified the presence of PHB in *Bacillus megaterium*. PHB can be accumulated up to 80% of the cell dry weight from various carbon sources by *Ralstonia eutropha* and near 90% in recombinant *Escherichia coli* (Du et al., 2012).

7.1.1   CHEMISTRY OF THE PHAS

PHAs are composed of 3-hydroxy fatty acid monomers, which form linear, head-to-tail polyester (Fig. 7.1). PHAs which are products of bacterial activity are most interesting among all of biodegradable plastics. PHAs are polymers of long molecules which are made of many small monomers that are linked together. These polymers are not soluble in the water and are biodegradable, have thermoplastic characteristics, and could be produced from recyclable carbon sources, for example, waste cooking oil. Thermoplastics are polymers that become flexible at a temperature higher than a specific temperature, and return to a solid state after cooling. Most of thermoplastics have a high molecular weight, and their chains correlate via intermolecular forces; this exclusivity causes thermoplastics to be molded again because the intermolecular interactions form again after cooling (Valappil et al., 2007). The monomers that produce PHA are 3-hydroxyalkanoates (3HA). An alkanoate is a free fatty acid that has a linear molecule structure containing hydrogen and carbon with a carboxyl group at one end. These monomers contain a hydroxyl group (OH) at the third carbon atom, making these beta or 3HA. By an ester bond, the hydroxyl groups in one monomer are linked to the carboxyl groups of other monomer (Jacquel et al., 2008). Figure 7.1 shows that the polyester link creates a molecule which has three carbon parts that are separated by oxygen atoms. The residue of the monomer becomes a side chain of the main backbone of the polymer.

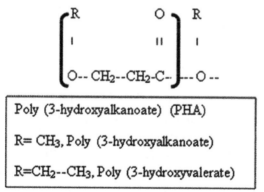

FIGURE 7.1   General structural formula of polyhydroxyalkanoate (PHA).

7.1.2 PHYSICAL PROPERTIES OF PHAS

The composition of the PHA directly affects the physical properties of the plastic. Polymers created from hydroxyoctanoate, including minimum of eight carbon monomers, are elastic. Long-side-chain polymers are very flexible and gummy. By feeding bacteria producing PHA with a convenient substance, a PHA polymer with specific suitable properties can be produced. PHAs can be classified into various groups as shown in Table 7.1. Among them, classifications based on monomer size and the type of polymer are the most common (Jantima et al., 2010).

TABLE 7.1 Classification of Microbial Biopolymers (PHAs) at Different Scales (Adapted from Luengo et al., 2003).

Factors of classification	Details on the production and properties
Biosynthetic source	Natural biopolymers (PHAs): produced naturally by bacteria strains, from common substrates, such as poly-3-hydroxybutyrate (P3HB)
	Semi-synthetic biopolymers (PHAs): the addition of unusual precursors, that is, 3-mercapto propionic acid to increase the biosynthesis of poly(3-hydroxybutyrate-co-3-mercaptopropionic)
Size of monomer (number of carbon atoms)	Short-chain-length PHAs: scl-PHA (3–5 carbon atom) Medium-chain-length PHAs: mcl-PHA (6–14 carbon atoms)
Number of different monomers in biopolymer (PHA)	Homo polymer: the polymerization begins with the linkage of a small molecule or monomer via ester bonds to decarboxylic group of the next monomer. Hetero polymer: Two or more different monomer units are linked together

7.1.3 BIODEGRADABILITY AND BIOCOMPATIBILITY OF PHAS

Biodegradability and biocompatibility are important characteristics of PHAs. These properties make it interesting in comparison with synthetic plastics. Having the similar properties to the thermoplastics and bestowed with the biodegradability, materials made from PHAs are expected to replace the traditional petroleum-based plastics (Du et al., 2012; Ida et al., 2006). Biodegradation depends on several factors, for example, environmental microbial activity and the revealed surface area, pH, temperature, moisture, and molecular weight. The development of copolymer production or blending PHAs with other monomers has widened their applications (Du et al., 2012). The natural base of the monomers also produces effect on PHA degradation in the environment. Copolymers which contain PHA monomer units can to be degraded faster than PHB or 3HB-*co*-3 HV copolymers (Anderson & Dawes, 1990).

Enzymes naturally can hydrolyze the biopolymers into their molecular building blocks, which are called hydroxyacids. These building blocks have served as a source of carbon for the growth of bacteria. Aerobic biodegradation of PHA produces carbon and water, and biodegradation of PHA at anaerobic conditions produces carbon dioxide and methane. Moreover, PHA is recyclable and degradable over a wide range with moisture levels at 55% and temperatures at a maximum of almost 60°C (Jawahar et al., 2012). Studies have shown that in such conditions, 85% of biopolymers (PHAs) were degraded in almost 7 weeks. It has been reported that PHA degrades in aquatic environments within 254 days at temperatures less than 6°C (Reddy et al., 2003). The molecular weights of PHA samples remained almost unchanged during the course of biodegradation. A number of microorganisms such as bacteria and fungi in soil, sludge, and sea water excrete extracellular PHA-degrading enzymes to hydrolyze the solid PHA into water-soluble oligomers and monomers, and subsequently utilize the resulting products as nutrients within cells. Figure 7.2 describes the life cycle of PHAs.

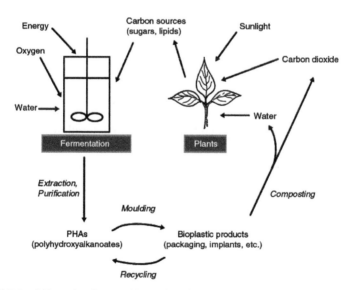

FIGURE 7.2   Life cycle of PHAs (Gross & Kalra, 2002).

PHAs are mainly produced from recyclable sources, thus they are not dependent on the availability of oils as feedstocks and also the production of energy for the biopolymer (PHA) generation processing depends on recyclable sources (Chen, 2009). Carbon dioxide (CO_2) has been distributed as one of the products of final mineralization of biopolymers (PHA), emanates via recycling source of carbon for the biosynthesis. The photosynthetic nomination of the liberated carbon dioxide through plants causes to generate recyclable carbon sources one more time (Joanne et al., 1996). During this process, the carbon sources amount in the synthesis and degradation of PHAs will be equivalenced. Hence, PHA does not participate in global warming (Jun Xu et al., 2006).

In some fields of application, PHAs are superlative in comparison with fuel plastics due to their biocompatibility. The most arbitrary biocompatibility of PHAs is enforced through natural occurrence of (R)-3-hydroxybutiric acid (3HB) (Steinbücchel et al., 1995; Zinna et al., 2001).

7.2   PRODUCER OF POLYHYDROXYALKANOATES

PHAs are synthesized by many living organisms. The main candidates for the large-scale production of PHAs are plants and bacteria. Plant cells can only cope with low yields (<10% (w/w) of dry weight) of PHA production. High levels (10–40% (w/w) of dry weight) of polymer inside the plant have a negative effect on the growth and development of the plant. In contrast, within bacteria, PHAs are accumulated to levels as high as 90% (w/w) of the dry cell mass. Accumulating PHAs is a natural way for bacteria to store carbon and energy, when nutrient supplies are imbalanced. These polyesters are accumulated when bacterial growth is limited by depletion of nitrogen, phosphorous, or oxygen and an excess amount of a carbon source is still present (Verlinden et al., 2007).

7.2.1   BACTERIA

Many species of bacteria, which are members of the family Halobactericeae of the Archaea, synthesize PHAs. The list of such microorganisms is growing and currently contains more than 300 organisms. *Ralstonia eutropha* (also known as *Cupriavidus necator* or *Alcaligenes eutrophus*) is among the bacteria that have been extensively studied for the production of PHAs. The chemical diversity of PHAs is large, of which the most well-known and widely produced form is PHB. The synthesis of

PHB is considered the simplest biosynthetic pathway. Figure 7.3 shows the PHAs biosynthetic pathways in *R. eutropha*; the process involves three enzymes: β-ketothiolase (phaA)/3-ketothiolase (bktB), acetoacetyl-CoA reductase (PhaB), and PHA synthase (phaC). In *R. eutropha*, two acetyl-CoA molecules are condensed to acetoacetyl-CoA by an enzyme β-ketothiolase (PhaA). Then, the reduction of acetoacetyl-CoA to (R)-3-hydroxybutyryl-CoA catalyzed by the acetoacetyl-CoA reductase (PhaB) is known as an NADPH-dependent reductase which produces the (R)-3-hydroxybutyryl-CoA monomers catalyzed by the last enzyme PHA synthase (phaC) responsible for polymerization of the monomers (Suriyamongkol et al., 2007). *R. eutropha* was also capable of producing the P(HB–HV) copolymer through alterations in the type and relative quantity of the carbon sources in the growth media such as addition of propionic acid or valeric acid in glucose media leads to the production of a random copolymer composed of HB and HV [P(HB–HV)] (Suriyamongkol et al., 2007). A few important other strains that were recently studied include: *Bacillus* spp., *Alcaligenes* spp., *Pseudomonas* spp., *E. coli*, *Rhizobium meliloti*, *Burkholderia sacchari*, and *Halomonas boliviensis*. Table 7.2 exhibits an overview of bacterial strains used to produce PHAs, including their corresponding initial carbon sources and produced (co)-polymers.

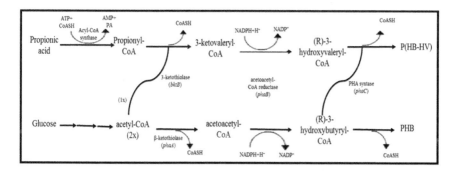

FIGURE 7.3 PHB and P(HB–HV) biosynthetic pathways.

TABLE 7.2 PHAs Production by Different Bacteria.

Bacteria	Carbon sources/feedstock	Yield	Reference
Alcaligenes eutrophus (now *Ralstonia eutropha*) H16 wild type and recombinant *R. eutropha* PHB-4	Olive oil, corn oil, palm oil	P(3HB), P(3HB-*co*-3HHx)	Fukui et al. (1998)
R. eutropha	Palm kernel oil, palmolein, crude palm oil, palm acid oil	PHB-4	Loo et al. (2005)
Gram-negative bacterium FLP1, *Burkholderia cepacia*	Palm oil mill effluent (POME)	P(3HB)	Alias and Tan (2005)
R. eutropha H16 wild-type and recombinant *R. eutropha* PHB-4 containing the *A. caviea*	Soybean oil	P(3HB) and P(3HB-*co*-3HHx)	Kahar et al. (2004)
Escherichia coli GCSC4401, GCSC6576 strains	Whey	Poly(3-hydroxybutyrate) (P(3HB))	Lee et al. (1997)
Methyl bacterium sp. ZP24	Cheese whey	P(3HB)	Yellore and Desai (1998)
R. eutropha	Whey-based basal medium + invert sugar + propionic acid	poly(3-hydroxybutyrate-*co*-3-hydroxyvalerate) (P(3HB-*co*-3HV))	Marangoni et al. (2002)
Sinorhizobium meliloti 41, *Hydrogenophaga pseudoflava* DSM 1034	Cheese whey	P(3HB)	Povolo and Casella (2003)
S. meliloti	Cheese whey	P(3HB)	Povolo and Casella (2003)
H. pseudoflava	Cheese whey		
Azotobacter vinelandii	Sucrose or cheese whey	P(3HB)	Dhanasekar and Viruthagiri (2005)
Pseudomonas cepacia	Lactose + D-xylose	P(3HB)	Young et al. (1994)
A. vinelandii	Beet molasses + valerate	P(3HB-*co*-3HV)	Page et al. (1992)
Bacillus sp., JMa5	Cane molasses + sucrose	P(3HB)	Wu et al. (2001)

TABLE 7.2 *(Continued)*

Bacteria	Carbon sources/feedstock	Yield	Reference
Pseudomonas spp.	P. cepacia	P(3HB)	Celik and Beyatli (2005)
	G13		
Rhizobium meliloti		P(3HB)	Mercan and Beyatli (2005)
Bacillus cereus M5	Sugar beet molasses as a carbon source	P(3HB)	Yilmaz and Beyatli (2005)
Methylo bacterium	Glycerol (50 %), casein peptone	P(3HB)	Bormann and Roth (1999)
Rhodesianum			
Pseudomonas oleovorans strain NRRL B-14682	Glycerol	Scl-PHA	Ashby et al. (2005)
P. oleovorans NRRL B-14682			
Pseudomonas resinovorans	Triacylglycerol	mcl-PHA	Cromwick et al. (1996)
Pseudomonas stutzeri 1317	Soybean oil	mcl-PHA	He et al. (1998)
P. corrugata	Soybean oil-based biodiesel	mcl-PHA	Ashby et al. (2004)
P. corrugata	Pure glycerol	mcl-PHA	Ashby et al. (2005)
Pseudomonas species	Soy molasses medium	mcl-PHA	Solaiman et al. (2006)
P. corrugata			
P. putida IPT046	Rice oil	PHA-3-hydoxydecanoate (3-HD)	Chee et al. (2010)
Pseudomonas aeruginosa IPT171	Rice oil	P(3HO)	Chee et al. (2010)
P. mosselii TO7	C6 fatty acid, gluconate, or fructose	3-HO	Chen et al. (2014)
P. mosselii TO7	Octanoic acid	P(3HO)	Chen et al. (2014)
P. mosselii TO7	Palm kernel oil	mcl-PHA	Chen et al. (2014)

7.2.2 EUKARYOTIC CELLS

The eukaryotic cell is another potential producer of PHAs. The production of bioplastic in bacteria is limited by its high cost compared to the costs associated with petroleum-derived plastics production. This aspect has been one of the driving forces in exploring eukaryotic systems, especially crops, as production hosts. Studies of PHA formation in yeast and insect cells can provide valuable information about how these pathways can be incorporated into plants (Suriyamongkol et al., 2007). Synthesis of PHB has been demonstrated in *Saccharomyces cerevisiae* by expressing the PHB synthase gene from *R. eutropha* (Leaf et al., 1996). However, PHB accumulation was very low (0.5% of cell dwt), possibly because of insufficient endogenous β-ketoacyl-CoA-thiolase and acetoacetyl-CoA reductase activities. Poirier et al. (2001) introduced a modified phaC1 gene from *P. aeruginosa* into *S. cerevisiae*. Other researchers also have explored the possibilities of changing monomer recombinant yeast cells to improve the production of PHAs (Marchesini et al., 2003; Zhang et al., 2006; Suriyamongkol et al., 2007).

7.2.3 TRANSGENIC PLANTS

PHA production in bacteria and yeast requires growth under sterile condition in a costly fermentation process with an external energy source such as electricity. In contrast, PHA production in plant systems is considerably less expensive because the system only relies on water, soil nutrients, atmospheric CO_2, and sunlight. In addition, a plant production system is much more environmental friendly. Plants use photosynthetically fixed CO_2 and water to generate the bioplastic, which after disposal is degraded back to CO_2 and water (Suriyamongkol et al., 2007). Synthesis of PHAs in crops is also an excellent way of increasing the value of the crops. Since starch and sugar are produced in plants at costs below the cost of commodity plastics, it might be possible to produce PHA at a similar low cost. Unlike the bacterial cell, the plant cell has different subcellular compartments in which PHA synthesis can be metabolically localized. As mentioned earlier, PHB is synthesized in bacteria from acetyl-CoA. This thioester is present in plant cells in the cytosol, plastids, mitochondria, and peroxisomes (Fig. 7.4). Therefore, it should be possible to produce PHB in any of these subcellular compartments (Hanley et al., 2000; Moire et al., 2003; Snell & Peoples, 2002; Suriyamongkol et al., 2007).

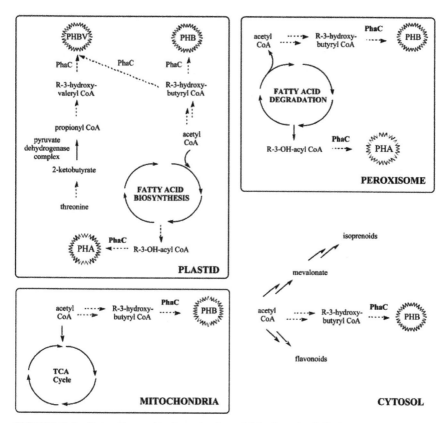

FIGURE 7.4 Formation and polymerization of 3-hydroxyacyl-CoA monomers in plant cells. Solid lines indicate native plant enzyme activities. Dashed lines indicate enzyme activities engineered into the plant for PHA formation. The following abbreviations are used: PhaC, PHA synthase; TCA cycle, tricarboxylic acid cycle.

7.3   CARBON SOURCES/FEEDSTOCK FOR PRODUCTION OF POLYHYDROXYALKANOATES

The utilization of bioprocesses due to the industrial production of biopolymers (PHA) could be regarded as undertaking for tolerable development. The cost of the raw materials as the first materials for producing PHA is an important factor that causes this matter. Thus, a permanent and useful solution method could be recognized during the consumption of a wide reach of waste materials and excess materials which could be upgraded to the function of feedstocks for biomediated generation of favorable final yield. These materials are basically generated and manufactured in agricultural and industrial fields which are completely related to agricultural industries (Braunegg et al., 1978; Khanna & Srivastava, 2005a,b). The simultaneous utilization of confined fossil sources for the generation of polymers stimulates overcoming worldwide difficulties, for example, the global warming and the greenhouse effect. It shows that these materials are employed only for short extent. Then, they will be mostly incinerated, increasing the atmospheric carbon dioxide (CO_2) concentration, taking part in the heating processes. During the incineration of fuel–polymers wastes, the chemical energy stored inside the fuel–plastics is recovered as thermal energy.

First materials cost and biopolymer (PHA) yield are the important factors that determine the overall cost for PHA production. Amongst the several nutrients in the fermentation medium in PHA production, the source of carbon has the significance of the total cost of substrate. Carbon sources, including carbohydrates $C_m(H_2O)_n$, hydrocarbons, oils, acids, and alcohols can be used by numerous bacteria (Yamane, 1993). The global region of the plant which will be built has a very important effect on choosing the suitable waste stream as a first material for bioprocessing targets. The waste streams utilizing for the generation of high amount yields increases the economics of these yields and also prepares the industry to dominate access matters (Sidik and Hori, 2008).

Generally, biopolymer accumulation is favored by sufficient disposal of a proper carbon source and a restricting stockpile with macro components, for example, nitrogen (N), phosphate (P), and dissolved oxygen or micro components, for example, potassium (K), magnesium (Mg), sodium cobalt (Sc), calcium (C), sulfate (S), iron (Fe), manganese (Mn), copper (Cu), and tin (Kim et al., 2002; Helm et al., 2008). Based on the scale of industry, phosphates and nitrogen limitation are important factors for change from equivalenced microbial growth to PHA agglomeration. Only during recent years, the influence of the microelements such as iron (Fe), sulfate (S), and potassium (K) as limiting growth factors on PHA quality was considered particularity and when potassium (K) limitation occurred, the generation of ultrahigh molecular masses of PHA manifested (Helm et al., 2008).

Many attempts have been undertaken to seek renewable sources of carbon for the production of PHAs. Since early 1980s, many companies have tried to produce various PHAs on pilot or industrial scales considering that petroleum prices would increase due to its exhaustion and people might be willing to use environmental-friendly non-petrochemical-based plastics, termed biodegradable plastics, green plastics, bioplastics, or ecoplastics (Mojaveryazdi et al., 2013). Various carbon sources such as whey (Nikel et al., 2006; Yellore, 1998; Koller et al., 2011), molasses (Page, 1992; Wu et al., 2001; Albuquerque et al., 2007; Chaijamrus & Udpuay, 2008), fats and vegetable oil (He et al., 1998; Loo et al., 2005; Fernández et al., 2005; Ntaikou et al., 2009; Chaudhry et al., 2011), wastewater (Martinez-Toledo et al., 1995; Ceyhan & Ozdemir, 2011), and lignocellulose raw materials (Keenan et al., 2006; Van-Thuoc et al., 2008) are also being considered as potential carbon sources for PHAs production. Figure 7.5 and Table 7.3 show the low-cost carbon sources or feedstocks to produce PHAs.

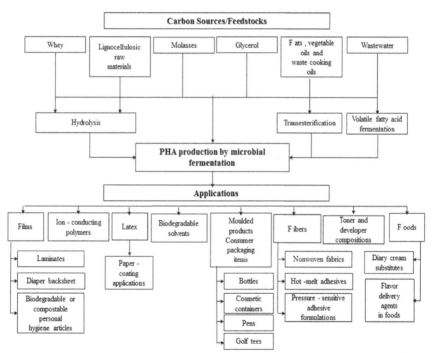

FIGURE 7.5 Carbon sources/feedstocks and applications of microbial PHA productions by different processing technologies (adapted from Du et al., 2012).

TABLE 7.3 Production of PHAs using Various Carbon Sources (Adapted from Du et al., 2012).

No.	Carbon sources	Strains	Types of PHAs
1.	**Molasses**		
	Sugar beet molasses	*Azotobacter vinelandii* UWD	PHB
	Sugarcane molasses	*Bacillus sp.* Jma5	PHB
	Molasses	*Bacterial consortium*	PHAs
	Sugarcane molasses	*Bacillus megaterium* ATCC 6748	PHB
		Bacillus megaterium BA-019	PHB
	Soy molasses	*Pseudomonas corrugata*	mcl-PHA
		Bacillus sp. CL1	PHAs
2.	**Whey and whey hydrolysates**		
	Hydrolyzed soy and malt	*Bacillus sp.* HF-1, HF-2	PHAs
	Hydrolyzed whey	*Ralstonia eutropha* DSM545	PHBV
		Pseudomonas hydrogenovora	PHBV
	Whey	Recombinant *E. coli*	PHB
		Thermus thermophiles HB8	PHAs
		Hydrogenophaga pseudoflava	PHAs
		Methylobacterium sp. ZP24	PHAs
		Methylobacterium sp. ZP24	PHB
		E. coli GCSC 6576	PHB
		E. coli GCSC 4401	PHB
3.	**Lignocellulosic raw materials**		
	Hemicellulosic fraction of poplar wood	*Pseudomonas pseudoava*	PHB
	Hemicellulosic hydrolysates	*Burkholderia cepacia* ATCC 17759	PHB
	Xylose with levulinic acid	*Burkholderia cepacia* ATCC 17759	PHBV
	Xylose; xylose with propionic acid	*Burkholderia cepacia* ATCC 17759	PHB; PHBV
	Xylose, glucose from sugarcane bagasse	*Burkholderia cepacia* IPT 048	PHB
		B. sacchari IPT 101	PHB
	Xylose and glucose	*E. coli* PTS *mutant*	PHAs
	Wheat bran hydrolysate	*Halomonas boliviensis* LC1	PHB
	Cellulose, in tequila bagasse	*Saccharophagus degradans* ATCC 43961	PHA
	Formic acid, acetic acid, furfural, and acid soluble lignin	*Ralstonia eutropha*	PHB

TABLE 7.3 *(Continued)*

No.	Carbon sources	Strains	Types of PHAs
4.	**Fats, vegetable oils, and waste cooking oils**		
	Unsaponifed olive oil	*Aeromonas caviae*	mcl-PHA
	Lard, butter oil, olive oil, coconut oil, soybean oil	*Pseudomonas aeruginosa* and *Pseudomonas resinovorans*	mcl-PHA
	Olive oil, corn oil, and palm oil	*Cupriavidus necator* (also known as *R. eutropha, A. eutropha*)	PHB
	Olive oil, corn oil, and palm oil	*P. putida*; *C. necator*	PHA
	Castor seed oil, coconut oil, mustard oil, cottonseed oil, groundnut oil, olive oil, and sesame oil	*Comamonas testosteroni*	The polymer contained HA units with 6–14 carbon atoms
	Lard and coconut oil	*Pseudomonas putida*	PHA
	Coconut oil and tallow	*Pseudomonas saccharophila*	mcl-PHA
	Palm kernel oil, palm olein, crude palm oil, and palm acid oil	*C. necator*	PHA
	Palm kernel oil	*C. necator*	P(3HB-Palm kernel oil *co*-3HV-*co*-3HHx)
		C. necator	PHA
	Soybean oil	*Pseudomonas stutzeri*	PHA
		C. necator H16 and *recombinant strains*	PHA
		C. necator mutants	PHA
	Jatropha	*Marine bacteria,* SM-P-3M	PHA
	Linseed oil	*Pseudomonas aeruginosa*	PHA
	Brassica carinata oil	*Pseudomonas aeruginosa*	PHA
	Waste cooking oil	*Pseudomonas aeruginosa*	PHA
	Spent palm oil	*Cupriavidus necator*	P(3HB-*co*-4HB)
	Waste vegetable oil	*Pseudomonas sp.* strain DR2	PHA
5.	**Wastewater**		
	Alpechin (wastewater from olive oil mill)	*Azotobacter chroococcum* H23	PHA
	Wastewater	*Enterobacter aerogenes* 12Bi	PHB

7.4   EXTRACTION, RECOVERY, DETECTION, AND QUANTIFICATION OF POLYHYDROXYALKANOATES

7.4.1   EXTRACTION OF PHAS

There are two common protocols used for PHA extraction from bacteria. The conventional PHAs extractions are:

1. Organic solvent extraction
2. Enzyme treatment.

For the organic solvent extraction, the PHA is extracted based on solubility of PHA in chloroform and insolubility in methanol. After harvest, lipids and other lipophilic components in the bacterial cells are removed by reflux in hot methanol followed by solubilization of PHA in warm chloroform. PHA from chloroform solvent can be recovered by solvent evaporation or precipitation by addition of methanol. Although highly purified PHA is obtained by this method, a large amount of hazardous solvent is needed to repeat the same process. Thus, this method is not environmental-friendly and unsuitable for mass production of bioplastic (Suriyamongkol et al., 2007).

The second protocol is designed to avoid the use of organic solvents. Bacterial cells are treated with a cocktail of enzymes (including proteases, nucleases, and lysozymes) and detergents to remove proteins, nucleic acids, and cell walls, leaving the PHA intact. In the large-scale production of PHA in crops, the extraction and purification of PHA from biomass is a critical factor for determining the practical feasibility of the technology. It is important that PHAs from transgenic plants can be extracted efficiently and easily, much like the extraction of endogenous compounds, such as starch, sucrose, and oil. Unlike extractions of bacteria, which are specifically intended for PHA production, there are other useful by-products that can also be extracted from harvested crops. Any extraction process from plant tissue should accommodate extraction of such compounds in unmodified form. The conventional methods used for extraction of low molecular weight lipids are not applicable for bioplastic produced in plant cells. Unlike separating vegetable oils from oilseeds, PHAs cannot be squeezed from the seeds by applying mechanical pressure. Solvent extraction is also difficult because the resulting polymer solution is extremely viscous, making the solution very difficult to work with. Also, the removal of solvent from the polymer is a slow and difficult process, and separations based on sedimentation are extremely slow. On a laboratory scale, the extraction of PHA from plant tissue has usually relied on the same method used to extract the polymer from bacteria (i.e., chloroform and methanol). Components of the recovered PHA are then analyzed using various analytical methods such as gas chromatography (GC)/mass spectrometry. There are no publications available in scientific literature regarding large-scale PHA extraction from plant tissues. There are methods, however, based on both solvent and nonsolvent procedures issued in the form of patents (Poirier et al., 2001; Suriyamongkol et al., 2007).

7.4.2   RECOVERY OF PHAS

After the fermentation processing, cells containing biopolymers (PHA) are separated by some other formal procedures, for example, centrifugation or filtration. Cells are interrupted to recover the polymers after the biomass harvesting (Ojumu et al., 2004). The first method that often has been used requires extraction with solvent. The chemical solvents include chloroform ($CHCl_3$), methylene chloride (CH_2Cl_2), propylene carbonate ($C_4H_6O_3$), and dichloroethane ($C_2H_4Cl_2$). The high volume of solvent needed makes this procedure economically undesirable, even after the recycling of solvents. Another method that has been created involves the use of sodium hypochlorite (NaClO) for the cellular material digestion without PHA. Use of sodium hypochlorite (NaClO) with chloroform ($CHCl_3$), significantly reduced degradation of PHA. Chloroform solves the P(3HB) by hypochlorite (NaClO), and maintains this polymer from degradation. By using hypochlorite treatment, polymer purity of more than 95% will be acquired (Preethi & Aravind, 2012). After synthesizing PHA by the microorganisms, the produced polymer will be rapidly hydrolyzed by increasing the activation of PHA-depolymerase enzyme. By decreasing the pH value to the pH 3 or pH 4, hydrolization of PHA can be achieved. In such pH conditions (3–4), the activity of (R)-(--)-3HB dehydrogenase enzyme would be almost zero, therefore the undesirable response of (R)-(--)-3HB to acetoacetate will be entirely blocked, and genuine form of (R)-(--)-3HB could be gathered (Koller et al., 2005).

7.4.3   DETECTION AND QUANTIFICATION OF PHAS

Intracellular PHA granules can be detected by a number of methods. The methods to detect, identify, and quantify the PHAs are as follows:

1. Staining dyes (Sudan black B, Nile blue A, and Nile red)
2. Electron microscopy (microscope, transmission electron microscopy (TEM))
3. UV spectrophotometer
4. Fourier transform-infrared spectroscopy analysis (FT-IR)
5. Gas chromatography
6. High performance liquid chromatography (HPLC)
7. Nuclear magnetic resonance (NMR)

Microbiologists have traditionally detected the presence of PHA granules in bacterial cells by staining with Sudan Black B. Native PHA granules can be stained with Sudan black B (Burdon, 1946) indicating that they are of a lipid nature (Kallio & Harrington, 1960). Sudan Black B staining was used in the microscopic observation for a possible accumulation of intracellular PHA produced by *Pseudomonas* (Fig. 7.6b). Electron microscopy was carried out with a Paxcam instrument. PHA granules can be observed by electron microscopy. Most of the PHA granules were placed close to the cytoplasmic membrane. Occasionally, PHA granules appear to be localized in the center of the cells. PHA accumulation was composed of varied monomer fractions ranging from 3-hydroxyhexanoate to 3-hydroxytetradecanoate. Figure 7.6a and b supposes that the PHA synthase is associated with the cytoplasmic membrane by hydrophobic interactions. This results in the formation of P(3HB) granules between the two layers of the phospholipid bilayer. The growing P(3HB) granule is then removed from the membrane and the other PHA-specific surface proteins attach to the surface of granules. Several studies are in agreement with this model of granule formation and reported that the emerging PHA granules tend to locate near the cell poles (Jendrossek & Gebauer, 2005).

PHA production and identification by UV spectrophotometer was estimated as previously described by Law and Slepecky (1961). PHA concentration was determined from an established standard graph in which the absorbance was plotted against the concentration of crotonic acid as a standard (235 nm). PHA granules extracted by the boiling chloroform method were used for measurement at a UV spectrum between 200 nm and 400 nm. To confirm the presence of PHA, the presence of a peak should be obtained between 230 nm and 240 nm (Santhanam & Sasidharan, 2010).

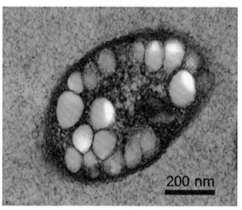

FIGURE 7.6   PHAs granules inside cytoplasm of the cell.

The amount of PHA in the sample can be determined by spectrophoto-metric assay. PHA polymer was extracted with chloroform and was exposed to evaporation of chloroform from the tube, and then 10 mL of sulfuric acid was added to the polymer and was heated at 100°C on a water bath for 10 min. By adding sulfuric acid (H_2SO_4) to the polymer, the polymer converts to crotonic acid and the color inside the glass tube changes to brown. Crotonic acid standard solution was prepared with different increasing concentrations which were 10–40 µg. Before the determination of PHA, the solution was diluted using sterile distilled water. This solution was diluted to the dilution factor of 10. As the initial solution contains concen-trated sulfuric acid and cannot be read by UV-spectrophotometer, dilution of the solution is needed. PHA in the extracted sample was confirmed by FTIR spectroscopy. The absorbance of the UV-spectrophotometer was set to 235 nm and the sample was transferred to silica cuvette and the absorbance reading of the sample was recorded (Selvakumar et al., 2011). Figure 7.7 shows the UV absorption spectrum which is a distinct absorp-tion peak around 230 nm. The sample containing PHA was digested in concentrated sulfuric acid and consequently diluted with 0.014 N H_2SO_4 (0.3–12 µg of sample per mL of final concentration). UV analysis of the product from sulfuric acid digestion of sample confirmed the presence of a single peak whose retention time was identical to that of crotonic acid. The UV spectrum of this product was identical to that of crotonic acid.

PHA granules which were extracted by the boiling chloroform method were used for measurement at a UV spectrum between 200 nm and 400 nm. PHA concentration was determined from an established standard graph in which the absorbance was plotted against the concentration of crotonic acid as a standard (235 nm). To confirm the presence of PHA, the presence of a peak should be obtained between 230 nm and 240 nm. The blue line shows crotonic acid concentration and the red line shows PHA concentration.

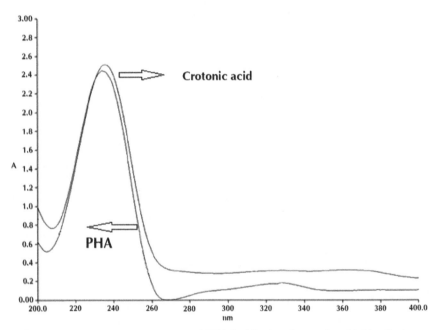

FIGURE 7.7   UV spectrophotometer of PHA (red line) and crotonic acid (blue line).

FT-IR spectroscopy (FT-IR Analysis) was also used to analyze the PHA production. The PHA extracted from the *P. oleovorans* was analyzed by FT-IR spectroscopy. To confirm the functional groups of the extracted polymer, it was used in the conditions of the 400–4000 cm^{-1} spectral range. Analysis exposed the presence of different conformations bands in the extracted PHA from mixed culture. The presence of absorption bands at 1722 cm^{-1} and 1279 cm^{-1} demonstrated the carbonyl bands C=O and C–O stretching ester in polymer. The bands at 1183 and 1134 cm^{-1} are characteristic of the asymmetric and the symmetric stretching of the C–O–C group, respectively (Santhanam & Sasidharan, 2010). In this study, the functional groups of the polymer PHA were confirmed as C=O groups by FT-IR spectroscopy. As is shown in Figure 7.8, the drastic absorption band located at approximately 1720 cm⁻¹ indicates the stretching vibration of the C=O groups in the PHA polyester. Accompanying bands of the C–O–C groups emerge in the spectral region from 1150 to 1300 cm⁻¹. One more absorption region is from 2800 to 3100 cm⁻¹, which corresponds to the stretching vibration of C–H bonds (Santhanam & Sasidharan, 2010). The absorption band at 2955.76 cm⁻¹ was assigned to an asymmetric methyl group. Asymmetric CH_2 of the lateral monomeric chains was assigned to the stretching vibration at 2925.98 cm⁻¹. Absorption at 1378.83 cm⁻¹ is assigned to terminal CH_3 groups. Series of absorption bands at 1166.87–619.39 cm⁻¹ were assigned to C–O and C–C stretching vibration in the amorphous phase (Gumel et al., 2012).

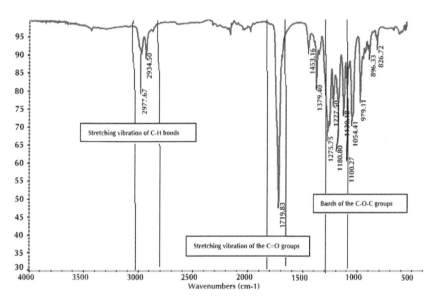

FIGURE 7.8   FT-IR spectrum of PHA produced by *Pseudomonas oleovorans*.

Intracellular PHA content also can be determined using GC analysis involving simultaneous solvent extraction and hydrolytic esterification of PHA. GC analysis with propanolysis in hydrochloric acid rather than acidic methanolysis in sulfuric acid was also reported. Quantification of microbial PHA using GC method is rapid, sensitive, reproducible, and requires only small amount of samples (5–10 mg) for the analysis. Other techniques of analysis such as two-dimensional fluorescence spectroscopy, flow cytometry, HPLC, ionic chromatography, microcalorimetric technique, and enzymatic determination were also described. For precise composition determination and structural elucidation of PHA, a variety of NMR spectroscopy techniques have also been applied and the most commonly used are proton (^{1}H) and carbon-13 (^{13}C) NMR (Sudesh & Doi, 2005; Sudesh, 2013).

7.5 APPLICATIONS OF POLYHYDROXYALKANOATES

7.5.1 OTHER TYPES OF PHAS AND APPLICATIONS

Different applications of PHA need various material natures of the bio polyesters. These characteristics could be obtained by tuning the composition of the biopolymer during the biosynthesis. The most prevalent delegate of biopolymers (PHA), to wit the PHB, shows a high value of crystallinity and exclusive process ability for these biomaterials (Tabassum et al., 2009). The miniature diversity between the disintegration temperature (almost 271°C) and also the melting point (almost 181°C) prepares a process ability for melt exclusion technology. This is changeable by disrupting the PHB matrix by interpolation of another building block like 3HV and the archival building blocks 4HB and 5HV. Thus, scl-PHA attributes the properties of thermoplastics, for example, polypropylene. These pathfinders participate in the generation cost, and also are toxic for the generation and production strains (Lee, 1996). Depending on the monomer composition, the characteristics of materials and the application of the biopolymers vary. Polyhydroxybutyrate, P(3HB), is a more famous biopolymer (PHA). Industrial applications of P(3HB) have been prevented and disturbed owing to its low thermal stability and inordinate friability and friability through storage. Copolymers of P (3HB-co-3HV), 3-hydroxyvalerate (3HV), 3-hydroxybutyrate (3HB), are more flexible and tougher than the P(3HB). These copolymers are useful to produce numerous products, such as coated paper, films, board, disposable food, fertilize bags, service ware and molded products, for example, bottles and razors and also is useful for biomedical applications (Lee, 1996).

Besides that, recently 4-hydroxybutyrate, P(4HB) has been discovered. The P(4HB) has more applications in biomedical usages (Martin et al.,

2005). P(4HB) also is used for tissue engineered heart valve and perma-nent ovine blood vessels. Likewise, a high molecular weight copolymer of 3HB and 4HB [P(3HB-*co*-4HB)] that contains 0–100 mol% of 4HB could be produced by *Comamonas acidovorans* strain under controlled degradation conditions (Saito & Doi, 1994), making them unique candi-dates for biomedical usages, for example, tissue engineering (Martin et al., 2005). The confinement of applications for biopolymers (PHA) are related to ordinary packaging first materials and also to embrace stuff items, stuff materials for agricultural purposes and in some cases for pharmaceutical and medical applications.

7.5.2 THE POSSIBLE APPLICATIONS OF PHA

Many products can be created. For example, manufacturing of bottles, different kinds of fibers and latex, and some yields of agricultural, commercial, or packaging interest, packaging films, compost bags can be done. Also, many kinds of carriers, biodegradable conveyors for long-term dosage of drugs, medicines, and herbicides, also some fertilizers, dispos-able things, for example, razors, vessels, diapers, or feminine pads; and some hygiene products can be made. Further, PHA has many medical appli-cations, for example, implants, urological stents, surgical pins, sutures, staples, neural and cardiovascular-tissue engineering, fracture fixation, therapy of narcolepsy and alcohol consuetude, drug delivery vehicles, cell microencapsulation, support of hypophyseal cells or as precursors of molecules with antirheumatic, analgesic, radiopotentiator, chemopre-ventive, antihelmintic, or antitumoral characteristics, and production of swabs, and wound dressing (Ren et al., 2005).

After utilization of these materials as plastic segments, biopolymers can be simply depolymerized and degraded to a beneficial source of optically genuine R-(--)-configured biofunctional hydroxyl acids which are more interesting as synthons for chiral value-added chemicals, for example, some kinds of vitamins, many antibiotics, some of pheromones, and aromatics in some cases in this field; acids also show major biological characteristics, for example, some activities such as antimicrobial and anti-viral activity (Ruth et al., 2007). Production of these compounds, relieving a market value higher than the polymer itself, using chemical procedures is complicated and difficult and it is not economical.

7.5.3   ECONOMICS OF PHAS PRODUCTION

The economic value of the raw materials is a significant factor in the economics of biopolymer (PHAs) production and is evaluated to a great vastness (more than 51% of the existing production costs). This insinuates high amounts of biopolymer as a quantity weight of dry cell and mentions the high amount of productivity for qualifications of gram of yield per each volume and time unit (Reddy et al., 2003). This is caused by the PHA agglomeration which starts at aerobic conditions and results in vast amount of decrements of the carbon substrate by intracellular aspiration. Accordingly, just a high quantity of less than 51% of the carbon source has been due to biomass and biopolymer production. The operation of waste materials raised to the inscription of first materials for PHA biosynthesis organizes a permanent strategy for cost-effective biopolymer production and supports industry to vanquish disposal matters (Tabassum et al., 2009).

One of the efforts for gathering waste lipids as a carbon source is to use cooking oil wastes, plant oils wastes, MBM lipids (meat and bone meal) or wastewater from oil-producing factories. Among all of these reasons, the triacylglycerides could be directly consumed as a source of carbon as after hydrolyzation process triacylglyceride changes to glycerol and free fatty acids, or after transesterification to glycerol. In a fermentation process, the costs of production of feedstocks can incorporate 71% of the total amount of PHA manufacturing costs for mass production of biochemicals (Hepner, 1996). The production crop of biopolymers (PHA) on the carbon source as a significant factor in the microbial generation of biopolymers (PHA) directly depends on the cellular PHA content, whilst the lower the amount of biopolymer in the cell the more source of carbon is wasted on extant biomass. Additionally, the PHA value is also an important factor in downstreaming processing costs and effects on the costs of the PHA production (Hazenberg & Witholt, 1997).

7.6   CONCLUSIONS

The natural secondary metabolites of polyhydroxyalkanoates (PHAs) produced by many species of microorganisms or transgenic plants are being considered as a replacement for conventional plastics (petroleum-derived plastics). The monomer composition, macromolecular structure, and physical chemical properties of PHA vary depending on the producer organism as well as the carbon source used for the growth. The unique property of PHAs is that these can be completely biodegraded within a year by a variety of microorganisms. There are variety of carbon sources and microorganisms used for production of PHAs. Then, PHAs are extracted either using solvent extraction or/and enzyme treatment. While for recovery process, PHAs are recovered and analyzed using various analytical methods such as gas chromatography/mass spectrometry. PHA granules are identified and characterized using different methods such as staining dyes, electron microscopy, GC, UV spectrophotometer, HPLC, NMR, and others. Meanwhile, PHAs can be further characterized, quantified, and graded into different quality based on its chemical structure, physical and chemical properties, process ability, and degree of biodegradability which has been one of the remarkable properties of PHAs, the so-called green biopolymer for a greener environment.

KEYWORDS

- polyhydroxyalkanoates
- 3-hydroxy fatty acid monomers
- polymers
- enzymes
- bacteria

REFERENCES

Albuquerque, M. G. E.; Eiroa, M.; Torres, C.; Nunes, B. R.; Reis, M. A. M. Strategies for the Development of a Side Stream Process for Polyhydroxyalkanoate (PHA) Production from Sugar Cane Molasses. *J. Biotechnol.* **2007,** *130,* 411–421.

Alias, Z.; Tan, I. K. Isolation of Palm Oil-utilising, Polyhydroxyalkanoate (PHA)-producing Bacteria by an Enrichment Technique. *Bioresour. Technol.* **2005,** *96*(11), 1229–1234.

Anderson, A. J.; Dawes, E. A. Occurrence, Metabolism, Metabolic Role, and Industrial Uses of Bacterial Polyhydroxyalkanoates. *Microbiol. Rev.* **1990,** *54,* 450–472.

Ashby, R. D.; Solaiman, D. K. Y.; Foglia, T. A. Bacterial Poly (hydroxyalkanoate) Polymer Production from the Biodiesel Co-product Stream. *J. Polym. Environ.* **2004,** *12,* 105–112.

Ashby, R. D.; Solaiman, D. K. Y.; Foglia, T. A. Synthesis of Short-/ Medium-Chain-Length poly(hydroxyalkanoate) Blends by Mixed Culture Fermentation of Glycerol. *Biomacromolecules* **2005**, *6*(2), 106–112.

Bormann, E. J.; Roth, M. The Production of Polyhydroxybutyrate by *Methylobacterium rhodesianum* and *Ralstonia eutropha* in Media Containing Glycerol and Casein Hydrolysates. *Biotechnol. Lett.* **1999**, *21*, 1059–1063.

Braunegg, G.; Sonnleitner, B.; Lafferty, R. M. A Rapid Gas Chromatographic Method for the Determination of Poly-β-hydroxybutyric Acid in Microbial Biomass. *Eur. J. Appl. Microbiol. Biotechnol.* **1978**, *6*, 29–37.

Burdon, K. L. Fatty Material in Bacteria and Fungi Revealed by Staining. *J. Bacteriol.* **1946**, *52*(6), 665–678.

Celik, G. Y.; Beyatli, Y. Determination of Poly-beta-hydroxy-butyrate (PHB) in Sugarbeet Molasses by *Pseudomonas cepacia* G13 strain. *Zuckerindustrie* **2005**, *130*, 201–203.

Ceyhan, N.; Ozdemir, G. Poly-hydroxybutyrate (PHB) Production from Domestic Wastewater Using *Enterobacter aerogenes* 12Bi strain. *Afr. J. Microbiol. Res.* **2011**, *5*, 690–702.

Chaijamrus, S.; Udpuay, N. Production and Characterization of Polyhydroxybutyrate from Molasses and Corn Steep Liquor Produced by *Bacillus megaterium* ATCC 6748. *Agric. Eng. Int.* **2008**.

Chaudhry, W.; Jamil, N.; Ali, I.; Ayaz, M.; Hasnain, S. Screening for Polyhydroxyalkanoate (PHA)-producing Bacterial Strains and Comparison of PHA Production from Various Inexpensive Carbon Sources. *Ann. Microbiol.* **2011**, *61*, 623–629.

Chee, J. Y.; Yoga, S. S.; Lau, N. S.; Ling, S. C.; Abed, R. M. M.; Sudesh, K. Bacterially Produced polyhydroxyalkanoate: Converting Renewable Resources into Bioplastics. In *Current Research, Technology and Education Topics in Applied Microbiology and Microbial Biotechnology*; Mendez-Vilas, A. Ed.; Formatex Research Center: Madrid; 2010, pp 1395–1404.

Chen. G. Q. A Polyhydroxyalkanoates Based Bio- and Materials Industry. *Chem. Soc. Rev.* **2009**, *38*, 2434–2446.

Cromwick, A.-M.; Foglia, T.; Lenz, R. W. The Microbial Production of Poly(hydroxyalkanoates) from Tallow. *Appl. Microbiol. Biotechnol.* **1996**, *46*, 464–469.

Dhanasekar, R.; Viruthagiri, T. Batch Kinetics and Modeling of Poly-β-Hydroxybutyrate Synthesis from *Azotobacter vinelandii* Using Different Carbon Sources. *India J. Chem. Technol.* **2005**, *12*, 322–326.

Du, C.; Sabirova, J.; Soetaert, W.; Lin, S. K. C. Polyhydroxyalkanoates Production from Low-cost Sustainable Raw Materials. *Curr. Chem. Biol.* **2012**, *6*(1), 14–25.

Fernández, D.; Rodríguez, E.; Bassas, M. Agro-industrial Oily Wastes as Substrates for PHA Production by the New Strain *Pseudomonas aeruginosa* NCIB 40045: Effect of Culture Conditions. *Biochem. Eng. J.* **2005**, *26*, 159–167.

Fukui, T.; Shiomi, N.; Doi, Y. Expression and Characterization of (R)-specific Enoyl Coenzyme A Hydratase Involved in Polyhydroxyalkanoate Biosynthesis by *Aeromonas caviae. J. Bacteriol.* **1998**, *180*, 667–673.

Gross, R. A.; Kalra, B. Biodegradable Polymers for the Environment. *Science* **2002**, *297*, 803–807.

Gumel, A. M.; Annuar, M. S. M.; Heidelberg, T. Biosynthesis and Characterization of Polyhydroxyalkanoates Copolymers Produced by *Pseudomonas putida* Isolated from Palm Oil Mill Effluent. *PLoS ONE* **2012**, *7*(9), e45214. DOI:10.1371/journal.pone.0045214.

Hanley, Z.; Slabas, T.; Elborough, K. M. The Use of Plant Biotechnology for the Production of Biodegradable Plastics. *Trends Plant Sci.* **2000**, *5*, 45–46.

Hazenberg, W.; Witholt, B. Efficient Production of Medium-chain-length poly(3-hydroxy-alkanoates) from Octane by *Pseudomonas oleovorans*: Economic Considerations. *Appl. Microbiol. Biotechnol.* **1997**, *3*, 588–596.

He, W.; Tian, W.; Zhang, G.; Chen, G.-Q.; Zhang, Z. Production of Novel Polyhydroxyalkanoates by *Pseudomonas stutzeri* 1317 from Glucose and Soybean Oil. *FEMS Microbiol. Lett.* **1998**, *169*, 45–49.

Helm, J.; Wendlandt, K.-D.; Jechorek, M.; Stottmeister, U. Potassium Deficiency Results in Accumulation of Ultra-high Molecular Weight Poly-beta-hydroxybutyrate in a Methane-utilizing Mixed Culture. *J. Appl. Microbiol.* **2008**, *105*, 1054–1061.

Hepner, L. Cost Analysis of Fermentation Processes. *Chimia* **1996**, *50*, 442–443.

Ida I. M.; Lee K. J.; Mohd A. M. N. Comparing the Hydrolytic Degradation of Poly-β-hydroxybutyrate, Poly-β-hydroxybutyrate-*co*-valerate and Its Blend with Cellulose Triacetate, *Malaysian Polym. J.* PRIM. **2006**, *1*(1), 39–46.

Jacquel, N.; Lo, C.-W.; Wei, Y.-H.; Wu, H.-S.; Shaw, S. W. Isolation and Purification of Bacterial Poly(3-hydroxyalkanoates). *Biochem. Eng. J.* **2008**, *39*(1), 15–27.

Jantima, T.; Cheng, T. I. X.; Alissara, R.; Takaya, H.; Koichi, Y.; Masahiko, S. Screening of PHA-Producing Bacteria 66 Using Biodiesel-Derived Waste Glycerol as a Sole Carbon Source. *J. Water Environ. Technol.* **2010**, *8*(4), 373–381.

Jawahar, N.; Mudaliar, N.; Senthilkumar P.; Narendrakumar; Samrot, A. V. Influence of Substrate Concentration in Accumulation Pattern of Poly(R) Hydroxyalkonoate in *Pseudomonas putida* SU-8. *Afr. J. Microbiol. Res.* **2012**, *6*(15), 3623–3630.

Jendrossek, D.; Gebauer, B. Assay of Poly(3-hydroxybutyrate) Depolymerase Activity and Product Determination. *Appl. Environ. Microbiol.* **2005**, *72*(9), 6094–6100.

Joanne, M.; Curley, B. H.; Robert W. L. Production of Poly(3-hydroxyalkanoates) Containing Aromatic Substituents by *Pseudomonas oleovorans*. *Macromolecules* **1996**, *29*(5), 1762–1766.

Jun Xu, X. Q.; Dai, J.; Cao, H.; Yang, M.; Zhang, J.; Xu, M. Isolation and Characterization of a *Pseudomonas oleovorans* Degrading the Chloroacetamide Herbicide Acetochlor. *Biodegradation* **2006**, *17*, 219–225.

Kahar, P.; Tsuge, T.; Taguchi, K.; Doi, Y. High Yield Production of Polyhydroxyalkanoates from Soybean Oil by *Ralstonia eutropha* and Its Recombinant Strain. *Polym. Degrad. Stab.* **2004**, *83*, 79–86.

Kallio, R.; Harrington, A. A. Sudanophilic Granules and Lipid of *Pseudomonas methanica*. *J. Bacteriol.* **1960**, *80*, 321–324

Keenan, T. M.; Tanenbaum, S. W.; Nakas, J. P. Microbial Formation of Polyhydroxyalkanoates from Forestry-based Substrates. *ACS Symp. Ser.* **2006**, *921*, 193–209.

Khanna, S.; Srivastava, A. K. Statistical Media Optimization Studies for Growth and PHB Production by *Ralstonia eutropha*. *Process Biochem.* **2005a**, *40*, 2173–2182.

Khanna, S.; Srivastava, A. K. Recent Advances in Microbial Polyhydroxyalkanoates. *Process Biochem.* **2005b**, *40*, 607–619.

Kim, M.; Ahn, Y. H.; Speece, R. E. Comparative Process Stability and Efficiency of Anaerobic Digestion; Mesophilic vs. Thermophilic. *Water Res.* **2002**, *36*(17), 4369–4385.

Koller, M.; Bona, R.; Hermann, C.; Horvat, P.; Martinz, J.; Neto, J.; Pereira, L.; Varila, P.; Braunegg, G. Biotechnological Production of Poly(3-hydroxybutyrate) with *Wautersia eutropha* by Application of Green Grass Juice and Silage Juice as Additional Complex Substrates. *Biocatal. Biotransform.* **2005**, *23*(5), 329.

Koller, M.; Hesse, P.; Salerno, A.; Reiterer, A.; Braunegg, G. A Viable Antibiotic Strategy against Microbial Contamination in Biotechnological Production of Polyhydroxyalkanoates from Surplus whey. *Biomass Bioenerg.* **2011**, *35*, 748–753.

Koutinas, A. A.; Lin, C. S. K.; Webb, C. Chapter 10—Developments in Cereal-Based Biorefineries (section 10.5 Polyhydroxyalkanoate (PHA) Production from Wheat). In *Advances in Biorefineries Biomass and Waste Supply Chain Exploitation*; Waldron, K. Ed.; Woodhead Publishing Limited: Cambridge; pp 316–319.

Law, J. H.; Slepecky, R. A. Assay of Polyhydroxybutyric Acid. *J. Bacteriol.* **1961**, *82*, 33–36.

Leaf, T. A.; Peterson, M. S.; Stoup, S. K.; Somers, D.; Srienc, F. *Saccharomyces cerevisiae* Expressing Bacterial PHB Synthase Produces Poly-3-Hydroxybutyrate. *Microbiology* **1996**, *142*, 1169–1180.

Lee, S. Y. Bacterial Polyhydroxyalkanoates. *Biotechnol. Bioeng.* **1996**, *49*, 1–14.

Lee, S. Y.; Middelberg, A. P. J.; Lee, Y. K. Poly(3-hydroxybutyrate) Production from Whey Using Recombinant *Escherichia coli*. *Biotechnol. Lett.* **1997**, *19*, 1033–1035.

Loo, C.-Y.; Lee, W.-H.; Tsuge, T.; Doi, Y.; Sudesh, K. Biosynthesis and Characterization of Poly(3-hydroxybutyrate-*co*-3-hydroxy-hexanoate) from Palm Oil Products in a *Wautersia eutropha* Mutant. *Biotechnol Lett.* **2005**, *27*, 1405–1410.

Luengo, J. M. G. B.; Sandoval, A.; Naharro, G.; Oliver, E. R. Bioplastics from Microorganisms. *Curr. Opin. Microbiol.* **2003**, *6*(3), 251–260.

Marangoni, C.; Furigo, A. Jr.; de Aragão, G. M. F. Production of Poly(3-hydroxybutyrate-*co*-3-hydroxyvalerate) by *Ralstonia eutropha* in Whey and Inverted Sugar with Propionic Acid Feeding. *Proc. Biochem.* **2002**, *38*, 137–141.

Marchesini, S.; Erard, N.; Glumoff, T.; Hiltunen, J. K.; Poirier, Y. Modification of the Monomer Composition of Polyhydroxyalkanoate Synthesized in *Saccharomyces cerevisiae* Expressing Variants of the β-oxidation-association Multifunctional Enzyme. *Appl. Environ. Microbiol.* **2003**, *69*, 6495–6499.

Martin, K. R. B.; Braunegg, G.; Hermann, C.; Horvat, P.; Kroutil, M.; Martinz, J.; Neto, J.; Pereira, L.; Varila. P. Production of Polyhydroxyalkanoates from Agricultural Waste and Surplus Materials. *Macromolecules* **2005**, *6*(2), 561–565.

Martinez-Toledo, M. V.; Gonzalez-Lopez, J.; Rodelas, B.; Pozo, C.; Salmeron, V. Production of Poly-β-hydroxybutyrate by *Azotobacter chroococcum* H23 in Chemically Defined Medium and Alpechin Medium. *J. Appl. Microbiol.* **1995**, *78*, 413–418.

Mercan, N.; Beyatli, Y. Production of Poly-β-hydroxybutyrate (PHB) by *Rhizobium meliloti*, *R. viciae* and *Bradyrhizobium japonicum* with Different Carbon and Nitrogen Sources, and Inexpensive Substrates. *Zuckerindustrie* **2005**, *130*, 410–415.

Moire, L.; Rezzonico, E.; Poirier, Y. Synthesis of Novel Biomaterials in Plants. *J. Plant Physiol.* **2003**, *160*, 831–839.

Mojaveryazdi, F. S.; Zain, N. A. M.; Rezania, S.; Kamyab, H. Production of Biodegradable Polymers (PHA) through Low Cost Carbon Sources: Green Chemistry. *Int. J. Chem. Environ. Eng.* **2013**, *4*(3), 183–188.

Nikel, P. I.; de Almeida, A.; Melillo, E. C.; Galvagno, M. A.; Pettinari, M. J. New Recombinant *Escherichia coli* Strain Tailored for the Production of Poly(3-Hydroxybutyrate) from Agroindustrial By-Products. *Appl. Environ. Microb.* **2006**, *72*, 3949–3954.

Ntaikou, I.; Kourmentza C.; Koutrouli E. C. Exploitation of Olive Oil Mill Wastewater for Combined Biohydrogen and Biopolymers Production. *Bioresour. Technol.* **2009**, *100*, 3724–3730.

Ojumu, T. V.; Yu, J.; Solomon, B. O. Production of Polyhydroxyalkanoates, a Bacterial Biodegradable Polymer. *Afr. J. Biotechnol.* **2004**, 3, 18–24.

Page, W. J.; Manchak, J.; Rudy, B. Formation of Poly(hydroxy-butyrate-*co*-hydroxyvalerate) by *Azotobacter vinelandii* UWD. *Appl. Environ. Microbiol.* **1992**, *58*, 2866–2873.

Page, W. J. Production of Polyhydroxyalkanoates by *Azotobacter vinelandii* UWD in Beet Molasses Culture. *FEMS Microbiol. Rev.* **1992**, *103*, 149–157.

Poirier, Y.; Erard, N.; Petétot, J. M. Synthesis of Polyhydroxyalkanoate in the Peroxisome of *Saccharomyces cerevisiae* by Using Intermediates of Fatty Acid β-Oxidation. *Appl. Environ. Microbiol.* **2001**, *67*, 5254–5260.

Poirier, Y. Polyhydroxyalknoate Synthesis in Plants as a Tool for Biotechnology and Basic Studies of Lipid Metabolism. *Prog. Lipid Res.* **2002**, *41*, 131–155.

Povolo, S.; Casella, S. Bacterial Production of PHA from Lactose and Cheese Whey Permeate. *Macromol. Symp.* **2003**, 197, 1–9.

Preethi, R. S. P.; Aravind. J. Microbial Production of Polyhydroxyalkanoate (PHA) Utilizing Fruit Waste as a Substrate. *Res. Biotechnol.* **2012**, *3*(1), 61–69.

Quan, J.; Tian, J. Circular Polymerase Extension Cloning of Complex Gene Libraries and Pathways. *PLoS ONE* **2009**, *4*(7), e6441.

Reddy, C. S; Ghai, R.; Rashmi, T.; Kalia V. C. Polyhydroxyalkanoates: An Overview. *Bioresour. Technol.* **2003**, 137–146

Ren, Q.; Grubelnik, A.; Hoerler, M.; Ruth, K.; Hartmann, R.; Felber, H.; Zinn, M. Bacterial Poly(hydroxyalkanoates) as a Source of Chiral Hydroxyalkanoic Acids. *Biomacromolecules* **2005**, *6*, 2290–2298.

Ruth, K.; Grubelnik, A.; Hartmann, R.; Egli, T.; Zinn, M.; Ren, Q. Efficient Production of (R)-3-hydroxycarboxylic Acids by Biotechnological Conversion of Polyhydroxyalkanoates and Their Purification. *Bio-macromolecules* **2007**, *8*, 279–286.

Saito, Y.; Doi, Y. Microbial Synthesis and Properties of Poly(3-hydroxybutyrate-*co*-4-hydroxybutyrate) in *Comamonas acidovorans*. *Int. J. Biol. Macromol.* **1994**, *16*, 99–104.

Salehizadeh, H.; Van Loosdrecht, M. C. Production of Polyhydroxyalkanoates by Mixed Culture: Recent Trends and Biotechnological Importance. *Biotechnol Adv.* **2004**, *22*(3), 261–279.

Santhanam, A.; Sasidharan, S. Microbial Production of Polyhydroxyalkanotes (PHA) from *Alcaligens* spp. and *Pseudomonas oleovorans* Using Different Carbon Sources. *Afr. J. Biotechnol.* **2010**, *9*, 3144–3150.

Selvakumar, K.; Srinivasan, G.; Baskar, V.; Madhan, R. Production and Isolation of Polyhydroxyalkanoates from *Haloarcula marismortui* MTCC 1596 Using Cost Effective Osmotic Lysis Methodology. *Eur. J. Exp. Biol.* **2011**, *1–3*, 180–187.

Sidik, M. H. U.; Hori, K. Palm Oil Utilization for the Simultaneous Production of Polyhydroxyalkanoates and Rhamnolipids by *Pseudomonas aeruginosa*. *Appl. Microbiol. Biotechnol.* **2008**, *78*(6), 955–961.

Snell, K. D.; Peoples, O. P. Polyhydroxyalkanoate Polymers and Their Production in Transgenic Plants. *Metab. Eng.* **2002**, *4*, 29–40.

Solaiman, D. K. Y.; Ashby, R. D.; Foglia, T.; Marmer, W. N. Conversion of Agricultural Feedstock and Coproducts into Poly(hydroxyalkanoates). *Appl. Microbiol. Biotechnol.* **2006,** *71,* 783–789.

Steinbücchel, A.; Aerts, K.; Babel, W.; Follner, C.; Liebergesell, M.; Madkour, M. H.; Mayer, F.; Pieper-Fürst, U.; Pries, A.; Valentin, H. E. Considerations on the Structure and Biochemistry of Bacterial Polyhydroxyalkanoic Acid Inclusions. *Can. J. Microbiol.* **1995,** *41*(13), 94–105.

Sudesh, K. Chapter 2—Bio-Based and Biodegradable Polymers (Section 2.4 Detection and Quantification of PHA). In *Polyhydroxyalkanoates from Palm Oil: Biodegradable Plastics*; Sudesh K. Ed.; Springer Heidelberg: New York, 2013, pp 17–19.

Sudesh, K.; Doi, Y. Polyhydroxyalkanoates. In *Handbook of Biodegradable Polymers*; Bastioli, C. Ed.; Rapra Technology Limited: UK, 2005, pp 219–256.

Suriyamongkol, P.; Weselake, R.; Narine, S.; Moloney, M.; Shah, S. Biotechnological Approaches for the Production of Polyhydroxyalkanoates in Microorganisms and Plants—A Review. *Biotechnol. Adv.* **2007,** *25,* 148–175.

Tabassum, M. S. A.-A.; Rahman, N. A. A.; Yee, P. L.; Shirai, Y.; Hassan, M. A. Fed-batch Production of P(3HB-*co*-3HV) Copolymer by *Comamonas* sp. EB 172 Using Mixed Organic Acids Under Dual Nutrient Limitation. *Eur. J. Sci. Res.* **2009,** *33*(3), 374–384.

Valappil, S. P.; Peiris, D.; Langley, G. J.; Herniman, J. M.; Boc-caccini, A. R.; Bucke, C.; Roy, I. Polyhydroxyalk-anoate (PHA) Biosynthesis from Structurally Unrelated Carbon Sources by a Newly Characterized *Bacillus* spp. *J. Biotechnol.* **2007,** *127,* 475–487.

Van-Thuoc, D.; Quillaguamn, J.; Mamo, G.; Mattiasson, B. Utilization of Agricultural Residues for Poly(3-hydroxybutyrate) Production by *Halomonas boliviensis* LC1. *J. Appl. Microbiol.* **2008,** *104,* 420–428.

Verlinden, R. A. J.; Hill, D. J.; Kenward, M. A.; Williams, C. D.; Radecka, I. Bacterial Synthesis of Biodegradable Polyhydroxyalkanoates. *J. Appl. Microbiol.* **2007,** *102,* 1437–1449.

Wu, Q.; Huang, H.; Hu, G.; Chen, J.; Ho, K. P.; Chen, G. Q. Production of Poly-3-hydroxybutrate by *Bacillus* sp. JMa5 Cultivated in Molasses Media. *Anton. van Leeuw.* **2001,** *80,* 111–118.

Yamane, T. Yield of Poly-D(−)-3-hydroxybutyrate from Various Carbon Sources—A Theoretical Study. *Biotechnol. Bioeng.* **1993,** *41,* 165–170.

Yellore, D. Production of Poly-3-hydroxybutyrate from Lactose and Whey by *Methylobacterium* sp. ZP24. *Lett. Appl. Microbiol.* **1998,** *26,* 391–394.

Yellore, V.; Desai, A. Production of Poly-3-hydroxybutyrate from Lactose and Whey by Methylobacterium sp. ZP24. *Lett. Appl. Microbiol.* **1998,** *26*(6), 391–394.

Yilmaz, M.; Beyatli, Y. Poly-β-hydroxybutyrate (PHB) Production by a *Bacillus cereus* M5 Strain in Sugarbeet Molasses. *Zuckerindustrie* **2005,** *130,* 109–112.

Young, F. K.; Kastner, J. R.; May, S. W. Microbial Production of Poly-β-hydroxybutyric Acid from D-xylose and Lactose by *Pseudomonas cepacia. Appl. Environ. Microbiol.* **1994,** *60,* 4195–4198.

Zhang, B.; Carlson, R.; Srienc, F. Engineering the Monomer Composition of Polyhydroxyalkanoates Synthesized in *Saccharomyces cerevisiae. Appl. Environ. Microbiol.* **2006,** *72,* 536–543.

Zinna, M.; Witholt, B.; Eglim, T. Occurrence, Synthesis and Medical Application of Bacterial Polyhydroxyalkanoate. *Adv. Drug Delivery Rev.* **2001,** *53,* 5–21.

CHAPTER 8

DIVERSE UTILIZATION OF PLANT-ORIGINATED SECONDARY METABOLITES

VASUDHA BANSAL[1]*, PAWAN KUMAR[2], SATISH K. TUTEJA[3], MOHAMMED WASIM SIDDIQUI[4], KAMLESH PRASAD[5], and R. S. SANGWAN[1]

[1]*Department of Chemical Engineering,Indian Institute of Technology, Hauz Khas, New Delhi 110016, India
[2]Center of Innovative and Applied Bioprocessing, Industrial Area, Mohali 160071, Panjab, India
[3]Academy of Scientific and Innovative Research, CSIR-Central Scientific Instruments Organisation, Sector-30, Chandigarh 160030, India
[4]Department of Food Science and Postharvest Technology, Bihar Agricultural University, Sabour, Bhagalpur 813210, Bihar, India
[5]Department of Food Engineering and Technology, Sant Longowal Institute of Engineering & Technology, Longowal 148106, Punjab, India
*Corresponding author, E-mail: vasu22bansal@gmail.com, vasudha@ciab.res.in.

CONTENTS

ABSTRACT

In this chapter, the potential utilization of secondary metabolites has been discussed in food industry, nutraceuticals, and cosmetic products, with an insight to the future research needs. This interdisciplinary discussion is helpful for the readers to apprehend the emerging applications of metabolites.

8.1   INTRODUCTION

Research during the past decade has provided the extensive scientific evidence regarding the health benefits of secondary bioactive compounds. This chapter elaborates the potential applications of secondary bioactive compounds as shown in Figure 8.1.. Here, the promising applications of secondary metabolites are discussed with recent investigations reported. Basically, secondary bioactive compounds are derived from various plants sources that play an essential role in many areas. The classification of secondary bioactive compounds can be allocated in various groups based on their chemical components and their structure, function, and biosynthesis. These are terpenoids, steroids, fatty acid-derived substances and polyketides, alkaloids, phenolic compounds, nonribosomal polypeptides, and enzyme cofactors.

However, on the basis of biosynthetic pathways, these plant secondary metabolites are categorized in three main groups: (1) phenolic compounds, (2) terpenes, and (3) nitrogen-containing compounds (Rea et al., 2010; Babbar et al., 2015). Furthermore, secondary bioactive compounds have also been economically important as flavors and fragrances, dyes and pigments, and food preservatives. The burgeoning commercial importance of secondary bioactive compounds has resulted in a great research and development interest in secondary metabolism, particularly the activities pertaining to altering the production of bioactive plant metabolites by means of tissue culture technology and metabolomics. This section presents information about secondary metabolites and their therapeutic applications (such as neurodegenerative diseases, cancer, diabetes, cardiovascular dysfunctions, inflammatory diseases and also ageing, etc.).

Secondary metabolites are reported to have anti-inflammatory, hepatoprotective, antihyperlipidemic, anticarcenogenic, and antimalignancy properties (Sawant, 2010). It was outlined that existence of flavonoids and several gallic acid derivatives render fruits, vegetables, herbs, and so on, an intense antioxidant activity (Anila & Vijayalakshmi, 2002). Their metabolic actions have shown a lipid-lowering activity because serum cholesterol, triglycerides (TGs), phospholipid, and low-density lipoproteins levels have been reduced following its dosage (Kim et al., 2005). The utilization of plant-based metabolites in nutraceuticals, food additives, and pharmaceuticals constitute them as a nutritional and therapeutic adjunct (Patel & Goyal, 2012). Further, addition of these functional components to the dietary supplements increases the bioavailability of micronutrients (such as iron, zinc, selenium) from the cereal-based foods (Gowri et al., 2001). Also, there has been a wide usage of bioactive metabolites in ayurvedic preparations (Scartezzini & Speroni, 2000; Deka et al., 2001),and others.

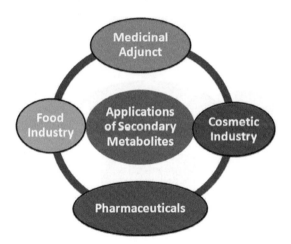

FIGURE 8.1   Diverse applications of secondary metabolites.

The color-rendering secondary metabolites like pigments, flavonoids are favored into products of jams, jellies, squashes, purees, dehydrated powders, and so on, in lieu of incrementing the color, aroma, and organoleptic acceptability. These compounds are usually acidic in nature which also plays the crucial role in providing shelf life to fruits and vegetables. Moreover, worldwide, food-based industries are continuously seeking novel products and creating inventive methods of processing from secondary metabolites (Puranik et al., 2012). It necessitates the development of edible products possessing superior functional components.

8.2 ROLE OF METABOLITES AS FOOD ADDITIVES IN AGROINDUSTRY

The biological active metabolites (such as polyphenols, flavonoids, flavo-nols, tannins, ascorbic acid, anthocyanins, carotenoids, sterols, saponins, glucosinolates, and volatile organic compounds) have been investigated for more than two decades for their physicochemical and nutritional char-acteristics (Dembitsky et al., 2011). Their role in supplementation of diets has enthralled the attention of food scientists and processing industries worldwide. The consumption of foods enriched with bioactive metabolites in the form of fruit and vegetable extracts, and herbal infusions are essen-tial and have been promoted as a supplement to everyday life. The poten-tial role of secondary metabolites in food industry is shown in Figure 8.2.

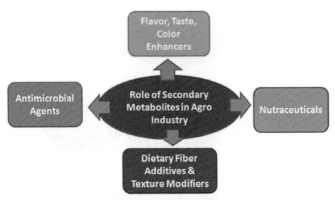

FIGURE 8.2 Schematic representation of utilization of secondary metabolites in food processing industry.

Owing to the awareness of the health benefits of metabolites as func-tional components, their processing is incrementing among the food processing industries. Further, improved techniques of preservation and transportation are adding to explore the large percentage of food in terms of the peels of fruits and vegetables, seeds, and their outer shells that are generated in huge numbers. Furthermore, these side products are much more richly endowed with the bioactive compounds relative to the pulp or juice (Ayala-Zavala et al., 2011). Due to these advantages, food indus-tries are utilizing them as food-quality enhancers in the form of flavoring compounds to enhance the taste and color of the other processed foods, as antioxidant agents, antimicrobials, colorants, texture modifiers, and forti-fiers for deficient processed products in order to cover-up their recommended level of micronutrients.

Therefore, these high valuable usables can be developed into nutra-ceuticals, natural food preservatives, taste enhancers, color-rendering compounds, supplementing cosmetic, and pharmaceutical industry as well (Gorinstein et al., 2011). In spite of their nutritional and organoleptic advantages, secondary metabolites also possess multivariant properties such as water-holding capacity, swelling capacity, gel forming which are very much essential for generation of certain products (Ayala-Zavala et al., 2011). The potential application of bioactive compounds can replace the chemical or synthetic compounds from the market which also further makes the consumer less concerned about the chemical formulas in the processed products (Janevska et al., 2010). The commercial opportunities of secondary metabolites in different dimensions of food industry are mentioned below:

8.2.1 FLAVORING AGENTS AS TASTE AND COLOR ENHANCERS

The synthetic additives have been rejected by the consumers and thus bioactive functional ingredients are being originated from natural sources. Flavoring properties are attributed to polyphenolics, tocopherols, vitamin C, and others, with particular adherence to low-molecular components (Schieber et al., 2001). Generally, plant volatiles are used as flavoring agents and many of them have fallen in the category of GRAS (generally regarded as safe). These bioproducts are largely available from the citrus industry (such as extracts of lemon, tangerine, oranges, etc.) (Conte et al., 2007). Moreover, the residues of citrus juices are the sources of dried pulp, molasses, essences, limonoids, flavonoids, and seed oil (Siliha et al., 2000). The major contributors of flavonoids from citrus varieties are hesperidin, narirutin, naringin, and eriocitrin (Schieber et al., 2001).

Similar to citrus fruits, others are mango, pineapple, guava, and banana. Due to the fatty acid (TG) profile of the mango seed, it has been employed in cocoa industry for supplementing the taste of cocoa butter (Arogba, 2000). The phenolics contributing to flavor in mango and pineapple are gallic acids, ellagic acid, and gallates. Similarly, anthocyanins existing in banana (such as delphinidin, cyanidin, peonidin, malvidin, and petunidin) have been used as color-providing pigments (González-Montelongo et al., 2010). Food products with good pigments are largely accepted by consumers and also command higher prices. Anthocyanins from grape pomace are one of the important colorants (Stintzing & Carle, 2004). Further, anthocyanins occur in wide range of red-to-blue color which are commonly found in berries (such as cherries, strawberries, plums, egg plants red cabbages, and radishes). Among this category, grape fruits are profuse with large phenolics in the form of hydroxycinnamic acid (p-coumaric, caffeic, sinapic, and ferulic acids), hydroxybenzoic acids (gallic acid, p-hydroxybenzoic acids (Georgiev et al., 2014)). These compounds are responsible for rendering color and flavor stability in wines. Wine is the most favored and desired product of nutritional benefit from grapes. Polyphenols such as flavonoids, anthocyanins, proanthocyanins, and stilbenes play the exploratory factors for wide range of food additives, and nutraceutical products from grapes (Georgiev et al., 2014). This wide range of products covers seed extracts, grape skin powder, and pomace powder. Recently, the Swiss company named "Mibelle Biochemistry" has launched the first commercial grape product for skin-care applications (PhytoCellTec™, 2013).

Seed powder from grapes is promoted as potential nutraceutical in the food industry (Greenspan et al., 2005). Flavor-producing flavonols present in the grapes are catechin, epicatechin, and proanthocyanidins.

8.2.2 ANTIMICROBIAL AGENTS

These days the lure for natural antimicrobials in the food industry has been increased due to the toxicological factors of the chemical preservatives. Essential oils present in the plant extracts cover the major share of antimicrobials. These essential oils have emerged as secondary metabolites by aromatic plants (Ayala-Zavala et al., 2011). Since ages, terpenes have been known for rendering antimicrobial activity. This is because terpenes have inhibitory effects against bacteria, fungi, and viruses. It may be due to the attack of terpenes on the respiratory cells (mitochondrion) of the organisms that prevent their proliferation (Burt, 2004). Similarly, the metabolites are used as preservatives. Lemon extracts were reported to be used in mozzarella cheese (Conte et al., 2007). Similarly, these compounds were found to be effective against meat and meat products. In addition to terpenes, phenolic compounds are also found to have optimum antimicrobial activity. This is due to the presence of hydroxyl groups in phenolic components (Dorman & Deans, 2000) and this hydroxylation results in increment of microbial toxicity among organisms. It was found that in the compounds like thymol, cymene, and menthol there is a presence of delocalized electrons around the hydroxyl groups. These electrons act as proton exchanger and reduce the pH around the cytoplasmic membrane, thus, interfering with the growth conditions of microbes. Consequently, reduction of ATP fosters the cell death (Ultee et al., 2002).

The increased interest of natural antimicrobials is also developed toward the extracts of herbs and spices in addition flavor compounds. These extracts have gained the popularity among the food-processing industries (Shan et al., 2007). The antimicrobial activity is attributed to the existence of phenolic compounds in the herbal extracts (Wu et al., 2006). A positive relationship has been observed between the concentration of bioactive compounds and toxicity action toward microbes. Resultantly, these compounds also enhance the shelf life of perishable food products. In addition to the phenolic and hydroxycinnamic acids in herbal extracts, essential volatile oils also have equal contributory antimicrobial effect (Holley & Patel, 2005). This category includes the oils of thyme, clove, cinnamon, sage, vanillin, organo, and citral which are found to be effective against bacteria. These plant extracts cover the phenolic compounds in the form of flavonoids, flavonols, glycosides, alkaloids, and polyacetylenes.

Likewise, the effect of electrons in the phenolics, sulfur element containing secondary metabolites also execute antimicrobial role, as the sulfur blocks the respiratory action of microbes which further resulted in the nonproliferation of cells. Metabolites present in onion and garlic possess these activities (Holley & Patel, 2005). The presence of allicin in garlic and onion is strongly antimicrobial. Similarly, among the spices, there are black pepper, cardamom, coriander, cumin, ginger, fenugreek, mint, curry leaves, paprika, and sesame which were found to have potent antimicrobial characteristic.

8.2.3 DIETARY FIBER ADDITIVES AND TEXTURE MODIFIERS

The components like functional foods, nutraceuticals, and designer foods have been usually employed for the non-nutrition contribution for attaining the health and medicinal health. Therefore, formulation of products with increased dietary fiber helps in preventing the chronic diseases and also leads to adequate consumption of associated bioactive compounds (Larrauri, 1999). On a commercial scale, formulated food products with high dietary fiber content are easily accessible. These products have the functional value and adequate content of dietary fiber is also available. Most of these fiber-rich products are usually generated from soluble and nonsoluble fiber. Soluble fibers are produced from cereal-based food products, fruits and vegetables. These fibers are easily incorporated into the high-fat content foods in order to dilute their ratio. Further, foods endowed with refined ingredients are often formulated with functional compounds along with dietary fiber. These products are proved to be an economic boon to all the three industries of food, cosmetic, and pharmaceutical (Ajila et al., 2010).

Consumers are very much inclined toward buying these formulated fiber-rich products from the food-processing industry, because this eliminates the fear of toxicity of synthetic compounds. Some of the products enriched in functional components of dietary fiber are given to the clinical patients of gut disease in order to strengthen their microflora inside their intestine (Palafox-Carlos et al., 2010). These are further recommended against the colon cancer and cardiovascular diseases. Further, mixing of these dietary fibers with the breakfast cereals leads to their fortification with micronutrients. The dietary fiber will directly provide the essential phytochemicals present in them through their incorporation in diets (Ayala-Zavala et al., 2011).

Recently, dietary fiber from orange was used as fat replacer in desserts like ice-cream (de Moraes Crizel et al., 2013). An ideal dietary fiber must have five essential characteristics as good flavor, color, odor, and balanced composition of bioactive compounds. The dietary fiber of citrus peels contains flavonoids, polyphenols, carotenoids, and vitamin C (Marín et al., 2007). In addition to replacing the fat compounds with healthy bioactive compounds; they will also increase the composition of antioxidants in these processed products.

Dietary fiber has numerous technological aspects due to their excellent properties in terms of water-holding capacity, gelling structure, enhancing viscosity, and also potential to be used in place of fat content (O'Shea et al., 2012). Owing to these advantages of metabolites, enriched dietary fiber can provide a balanced processed food with all taste, texture, and health. Further, these fiber-rich processed products are inexpensive, low-caloric, taste enhancers, with improved property of food oxidative stability and emulsion. Presently, all the food industries are adding these phytochemical-rich dietary fibers to confectionary products in the form of bakeries, deserts, beverages, cereals like pasta, corns, and lastly meat products as well (Elleuch et al., 2011). This is because a large number of consumers are inclined in buying these foods but reluctant due to their high fat and synthetic compounds content. But fortification and dilution of fat and synthetic compounds with natural compounds have widened their choices and buying capacity.

8.3   ROLE OF METABOLITES AS A MEDICINAL ADJUNCT

Secondary metabolites are reliable and effective as drugs, flavor and fragrances, dye, pigments, pesticides, and food additives (as shown in Fig. 8.3) and are less expensive.

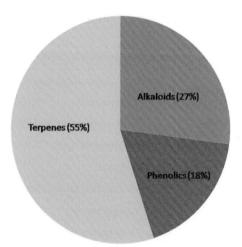

FIGURE 8.3   Pie chart representing the major groups of plant secondary metabolites according to Croteau et al. (2000), based on their numbers and diversity, due to which terpenes offer much potential in an array of industrial and medicinal applications.

Secondary metabolites are considered as end products of the primary metabolism and do not directly participate in metabolic processes; these include alkaloids, phenolics, tannins, lignins, resins steroids, and others.

8.3.1 ALKALOIDS

Alkaloids are secondary metabolites having diverse structures, primarily consisting of nitrogen and are widely used for pharmacological effect (Ahmed et al., 1986). Morphine was the first alkaloid to be found (Blakemore & White, 2002) from the plant *Papaver somniferum*, used as an analgesic and cough suppressant. Caffeine is the most lovable and famous known alkaloid. Caffeine is used as a CNS (central nervous system) stimulant and it comes from: cocoa, coffee, and tea. The seedlings of the coffee plants have a huge concentration of caffeine. The high concentration is toxic for insects and protects the plant from insects. In humans, caffeine is used to reduce the risk of diabetes and heart disease (Van Dam et al., 2006). Atropine is the medically important alkaloid commonly used as smooth muscle relaxant produced by plant commonly known as deadly night-shade (*Atropa belladonna*). It blocks the neurotransmitter acetylcholine (Cuthbert, 1963). Atropine is commonly used to dilate the pupil during eye and retina examinations (Bedrossian, 1979). Atropine also relieves nasal congestion and serves as an antidote to nerve gas (Geller et al., 2003) and insecticide-poisoning management (Eddleston et al., 2008) (Table 8.1).

TABLE 8.1   Enumeration of medicinal applications of secondary metabolites.

Atropine	Pupil dilation, ophthalmic applications
Caffeine	CNS stimulant
Codeine	Cough suppressant, bronchodilator, analgesic
Emetine	Antiprotozoal agent
Morphine	Analgesic
Quinine	Antimalarial, antipyretics
Reserpine	Antihypertensive
Vincristine	Anticancer, antitumor, anticarcinogenic
Vinblastine	Anticancer, antitumor, anticarcinogenic

The structure–activity relationship (SAR) is the relationship between the three-dimensional (3D) structure of a molecule and its biological activities which are used to enhance the medicinal primary effect of numerous synthetic and semi-synthetic modified drugs of the alkaloids and overcome the unwanted side effects. For example, naloxone, an opioid-receptor antagonist, is the derivative of the baine which is present in opium (Dewick, 2002). The preparations of plants containing alkaloids and their extracts, for example, cocaine, caffeine, and cathinone, are the stimulants of the CNS and are used as psychoactive substances (Manfred, 2002). Mescaline and various indole alkaloids (such as psilocybin, dimethyltryptamine, and ibogaine) have hallucinogenic effect (Veselovskaya et al., 2000a,d; Cordell, 2001). Cocaine is an example of alkaloid medicinally effective as anesthetic. Cocaine has long been used to alleviate hunger. However, cocaine derivatives are very dangerous when habitually used and can be deadly (Mack et al., 2010).

The other potent medicinally important alkaloids are the morphine and codeine used as strong narcotic pain killers and has analgesic property (Veselovskaya et al., 2000b). The ephedrine and pseudoephedrine alkaloid are the precursors for semi-synthetic psychoactive drugs like methcathinone and methamphetamine (Veselovskaya et al., 2000c). The baine is extremely useful in the synthesis of many analgesic preparations such as oxycodone. The quinine, derived from the bark of Cinchona tree, is used to reduce fever (Greenwood, 1992). The quinine is also effective against malaria that offered fewer side effects (Rahman & Mohammad, 2014). Cinchona bark also produces quinidine, which is effective for cardiovascular disorders particularly used to control abnormalities of heart rhythm such as fibrillation and heart blockage (Schwaab et al., 2009).

Vincaleukoblastine and vincristine, derived from *Catharanthus roseus*, are used effectively for the treatment of white-blood-cell cancers. Vincaleukoblastine is especially useful against lymphoma (cancer of the lymph glands), while vincristine is used against leukemia (Lucas et al., 2010). Pilocarpine, another alkaloid, is a very useful drug in ophthalmology for glaucoma treatment. This drug stimulates the drainage of excess fluid from the eyeball, relieving the high pressure in the eye caused by glaucoma (Brinchmann & Anmarkrud, 1979). The reserpine extracted from the genus *Raauwolfia* is useful to treat patients suffering from blood pressure by depleting neurotransmitter norepinephrine responsible for high blood pressure by arteries contraction (Garattini et al., 1961).

8.3.2 TERPENOIDS

The terpenoids are the secondary metabolites from plant origin and also synthesized by other organisms, such as bacteria and yeast. Terpenoids are made up of isoprenoid building blocks and based on the number of isoprenoid units, terpenoids are classified into several classes, such as monoterpenes, diterpenes, triterpenes, and tetraterpenes (Sacchettini & Poulter, 1997; Peñuelas & Munné-Bosch, 2005; Withers & Keasling, 2007; Rabi & Bishayee, 2009; Ashour & Wink, 2010).

The terpenoids are used as anticarcinogenic agents and illustrate cytotoxic effect against a variety of tumor and cancer cells. The terpenoids show other significant pharmacological activities, such as antiviral, antibacterial, antimalarial, anti-inflammatory, antiseptics, expectorants, gastrointestinal drugs, pain relievers, inhibition of cholesterol syntheses, other than anticancer activities (Mahato & Sen, 1997). Cannabis also acts as a non-psychoactive anti-inflammatory (Costa et al., 2007). The terpene is the active constituent of many herbs and spices. Terepene is an FDA-approved food additive, making it the first dietary cannabinoid and is also responsible for the spiciness of black pepper.

Another example is alpha-pinene, an organic compound found in the oils of rosemary and various species of pine trees. Pinene is medicinally effective and used to enhance mental focus and energy. Pinene is also useful as an antioxidant, bronchodilator, expectorant, anticancer, and a topical antiseptic (Aydin et al., 2013). Other terpenes such as limonene derived from citrus plants like oranges and lemons are antibacterial and antidepressants (Sun, 2007). Limonene also reflects relaxing effects, anticarcinogenic (Elson et al., 1988), and antifungal properties (Knoblocha et al., 1989). Next is taxol, which has become important in the medical field due to anti-carcinogenic effect to treat ovarian and breast cancer (Runowicz et al., 1993).

Myrcene is another abundant terpene and possesses muscle-relaxing, antidepressant, anti-inflammatory, and analgesic effects (Lorenzetti et al., 1991). Myrcene is also a very important chemical in the cosmetic industry because of its highly pleasant odor. Myrcene can be used for skin issues, respiratory disorders and as antiseptics. Camphor and turpentine, most commonly used terpenes for the treatment of respiratory disease, are used as diuretic, useful against gastrointestinal spasm, antiseptic and is also used in dentistry. Their fragrances are used in aromatherapy and effective against bacteria. Due to anesthetic properties, the ointments consist of camphor and turpentine and used to relieve pain and itching. The menthol is used to reduce flatulence and indigestion. Camphor and turpentine depress the CNS and can lead to respiratory disorder as well as nausea, vomiting, and seizures, which are typical symptoms of terpene poisoning. Other side effects include hallucinogenic effects and might lead to drug abuse. Essential oils give plants their essence and fragrance for aromatherapy and medicine. In some plants, the scent is used to detect herbivores and protect the plant from dangerous pathogens. Essential oils are useful to improve the mood, focus, and mental functioning. The largest among terepenoids is rubber consisting of 400 isoprene units. Rubber is obtained from latex of *Hevea brasilienis*. The uses for rubber are numerous mainly for shoes and textile industries, erasers, tires, gloves, and so on.

8.3.3   PHENOLICS

The phenol has recently become potential secondary metabolite due to intriguing beneficial effects on health. The phytochemical antioxidants and phenolic compounds are the secondary metabolites in plants, medicinal herbs, and dietary plants such as phenolic acids, flavonoids tannins, stilbenes, curcuminoids, coumarins, lignans, quinones, and carotenoids (Cotelle, 2001; Apak et al., 2007). Various bioactivities of phenolic compounds include antioxidant, anti-inflammatory effects and anticarcinogenic, antimutagenic effect (Huang & Ferraro, 2009). The flavonoids are the important group of secondary metabolite in phenol and further classified into the three main groups: anthocyanins, flavones, and flavonols (Narayana et al., 2001). Anthocyanins range in color from red to blue and purple and most commonly found in grapes, berries and have a wide range of health benefits. Anthocyanins show effective medicinal properties toward heart disease, diabetes, slow down aging process, and also show anticarcinogenic activity (Kim et al., 1993). The next two groups have white or yellow pigments. They are called flavones and flavonols. The salicylic acid is a kind of phenol that reflects analgesic properties, used in numerous skincare products to treat acne, large pores and dermatitis and also effective against hyperthermia. It has been used to effectively treat aches and fevers (Kim et al., 1993; Harborne & Williams, 2000). Natural phenolic compounds are used as anticarcinogenic agents and play an important role in the cancer prevention.

The phenolic compounds also show anticancer activities. The isolated polyphenols from different plants have been studied in a number of cancer cell lines at different stages of cancer growth and significantly show anticarcinogenic property (Damianaki et al., 2000; Zhang et al., 2008).

8.4 THERAPEUTIC APPLICATIONS OF SECONDARY METABOLITES

In present literature, there are many reports and review papers which are presented to show the therapeutic applications of secondary bioactive compounds (Kris-Etherton et al., 2004; Holdt & Kraan, 2011; Babbar et al., 2015). Their potential for reducing the disease risk has been observed such as coronary heart disease and chemopreventive properties in various forms of cancer. Several secondary bioactive compound studies related to the consumption of plant have shown a relatively low incidence of various type of cancer such as breast, colon, and prostate.

Plant foods contain a wide variety of bioactive phytochemical compounds with many potential bioactivities that may act as cancer preventive agents (Birt et al., 2001). There are many possible pharmacological mechanisms of bioactive phytochemical compounds by which intake of useful vegetables and fruits may associate with reduced cancer rates. Flavonoids and isoflavonoids are especially promising candidates found in numerous plants and are associated to play a prominent role in cancer prevention (Kale et al., 2008; Knekt et al., 2002). The different derivatives such as common phenylchromanone and hydroxyl structures of flavonoids are listed in Table 8.2 (Birt et al., 2001).

In recent literature, there has been a tremendous increase in the number of studies on flavonoids and isoflavonoids, which pay special attention to their abilities to inhibit the cell cycle, cell proliferation, and oxidative stress, and to induce detoxification enzymes, apoptosis, and the immune system (Birt et al., 2001). In the view of growing need and major interest in the pharmacological action, epidemiology, anticarcinogenic activity, bioavailability, and potential mechanisms of action, these secondary bioactive compounds are need to be focused. The number of mechanisms are followed by these bioactive compounds such as to prevent the carcinogens from reacting with cells at the initiation stage and block the promotion stage by inhibiting ornithine decarboxylase synthesis (Wattenberg, 1975). In future, knowledge of these secondary bioactive compounds will facilitate development of new strategies and approaches specifically as cancer preventive agents.

TABLE 8.2 Common Phenylchromanone and Hydroxyl Structures of Flavonoids (Birt et al., 2001).

Structural formula	Representative flavonoids	Substitutions					
		5	6	7	3'	4'	5'
Flavanone	Eriodictyol	OH	H	OH	OH	OH	H
	Hesperitin	OH	H	OH	OH	OMe	H
	Naringenin	OH	H	OH	H	OH	H
Flavanol	Catechin	OH	H	OH	OH	OH	H
	Gallocatechin	OH	H	OH	OH	OH	OH
Flavone	Apigenin	OH	H	OH	H	OH	H
	Chrysin	H	H	OH	H	H	H
	Luteolin	OH	H	OH	OH	OH	H
Flavonol	Kampherol	OH	H	OH	H	OH	H
	Myricetin	OH	H	OH	OH	OH	OH
	Quercetin	OH	H	OH	OH	OH	H
Flavanonol	Taxifolin	OH	H	OH	OH	OH	H
Isoflavone	Daidzein	H	H	OH	H	OH	H
	Genistein	OH	H	OH	H	OH	H
	Glycitein	OH	OMe	OH	H	OH	H
	Formononetin	H	H	OH	H	OMe	H

8.5   APPLICATION OF SECONDARY METABOLITES IN COSMETIC INDUSTRY

The presence of secondary metabolites in plants is also employed for cosmetic applications in practical world. In present literature, there are many reports, review papers, and books focusing on their potential cosmetic purposes (Kole et al., 2005; Schmidt et al., 2007; Parvez et al., 2007; Chuarienthong et al., 2010; Mukherjee et al., 2011). In this series, Kole et al. (2005) reported the cosmetic potential applications and they are generally classified depending upon the nature of their constituents as carotenoids, flavonoids, and polyphenols. In this review article, some Indian medicinal plants have been studied in detail for their cosmetics applications and are listed in Table 8.3. It has been showing the huge potential applications of metabolites against irregular dryness, dark/light pigmentation, sallowness, severe atrophy, telangiectasia, premalignant lesions, laxity, leathery appearance, and deep wrinkling. Moreover, the bioactive compounds are utilized for several synthetic skincare cosmetics existing in the market to treat premature aging and the most common adverse reactions of those include allergic contact dermatitis, irritant contact dermatitis, phototoxic and photo-allergic reactions (Kole et al., 2005; Mukherjee et al., 2011).

Interestingly, Chuarienthong et al. (2010) reported the herbal flavonoids extracted from the herbal miginkgo (*Ginkgo biloba*) a mixture of tea, rooibos (*Camellia sinensis* and *Aspalathus linearis*), and soybean (*Glycine soja*) for the herbal antiwrinkle cosmetics. This herbal formulation increased skin moisturization (27.88%), smoothness (4.32%), reduced roughness (0.4%), and wrinkles (4.63%), whereas the formula containing (*Ginkgo biloba*) a mixture of tea, and rooibos (*C. sinensis* and *A. linearis*) showed the best efficacy on wrinkle reduction (9.9%). Finally, study concluded that the herbal flavonoids from the tea and rooibos, gingko significantly improved for skin moisturization ($P = 0.05$) applications (Chuarienthong et al., 2010). In another interesting study, Parvez et al. (2007) reported the importance of tyrosinase, its biochemical characteristics, types of inhibitions, and sought new potent food and cosmetics applications. Likewise, number of investigation studies have been conducted on different models such as lab or animals models; however, there is still more research and development to be done to improve their safety and effectiveness in various applications.

TABLE 8.3 Indian Medicinal Plants and Their Part with Active Constituents Class for Their Cosmetics Applications (Kole et al., 2005).

S. No.	Name of the plant	Common name	Part used	Active constituents Class	Uses
1	Acacia concinna DC.	Shikakai	Pods	Saponin, sugars	Shampoo, soaps
2	Acorus calamus Linn.	Sweet flag	Rhizome	Monoterpenes, beta-asarone	Aromatic, dusting powders, skin lotions
3	Allium sativum Linn.	Garlic	Bulbs	Allicin and adenosine	Skin healing
4	Azadirachta indica A. Juss.	Neem	Leaves	Limonoids, tetranortriterpenoids	Toothpastes, soaps, shampoo
5	Cereus grandiflorus Mill.	Cactus	Leaf	Saponins, saccharides	Moisturizing tightening of skin
6	Cichorium intybus Linn.	Chicory	Seed	Sesquiterpene, lactones	Skin of blemishes
7	Citrullus vulgaris Schrad.	Water melon	Fruit	Saccharides, carotenoids, tannins	Sebum secretion
8	Citrus medica Linn.	Lemon	Fruit	Flavonoids, triterpenoids	Whitening, astringent depigmentation
9	Crocus sativus Linn.	Saffron	Stigma	Safranal, carotenoid	Post bath massage
10	Cucumis sativus Linn.	Cucumber	Fruit	Cucurbitacins	Moisturizing
11	Foeniculum vulgare Mill	Fennel	Fruit	Saponins, saccharides, flavonoids	Deodorant
12	Lawsonia alba Lam.	Henna	Leaves	Hanno-tannic acid, glucoside	Shampoo
13	Malus pumila Miller	Apple	Fruit	Saccharides, flavonoids, triterpenoids	Moisturizing, Anti-aging
14	Matricaria chamomilla Linn.	Chamomile	Flowers	Alpha-bisabolol, choline	Hair tonic
15	Mentha arvensis Linn.	Mint	Whole plant	Triterpenoids	Anti-perspiration
16	Panax ginseng Mey.	Ginseng	Root	Saponins, saccharides, triterpenoids	Hair-strengthening preparations
17.	Portulaca oleracea Linn.	Purslane	Whole plant	Saccharides, triterpenoids, tannins, saponins	Hair growth promotion, Moisturizing, anti-dandruff

TABLE 8.3 (Continued)

S. No.	Name of the plant	Common name	Part used	Active constituents Class	Uses
18	*Prunus armeniace* Linn.	Peach	Fruit	Saccharides, saponins, triterpenoids	Anti-aging creams Anti-kerati
19	*Pterocarpus santalinus* Linn.f.	Red sandalwood	Bark	Santalins	Skin creams
20	*Pueraria lobata* Ohwi	Arrot root Kudzu	Root	Flavonoids (isoflavones), saccharides, saponins	Astringent lotions
21	*Rheum coreanum* Nakai	Rhubarb	Root	Flavonoids, triterpenoids	UV absorption, free radical scavenger
22	*Santalum album* Linn	Sandal wood	Bark, wood	Sesquiterpenes, sesquiterpenols,	Skin lotions
23	*Triticum aestivum* Linn.	Wheat germ	Germ	Vitamin E, gliadin and glutenin	Skin lotions

8.6 CONCLUSION

Present and establishing secondary bioactive compounds and their therapeutical applications will reduce the risk of many diseases, including chronic diseases such as cancer, cardiovascular disease, and many other crude diseases. In addition, research and development on number of secondary bioactive compounds and the diversity of likely biological effects, numerous and diverse experimental approaches must be taken to increase our cognizance of the biology of bioactive compounds. Actually, the recognition of the bioactive complexity of biology, sophisticated experimental designs, and analytical methodologies must be employed to advance the field. Finally, we can say that more understanding of bioactive compounds in biological system will provide the scientific basis for future efforts to use biotechnology to modify/fortify foods and food components as a means to improve therapeutically.

KEYWORDS

- secondary bioactive compounds
- biosynthetic pathways
- metabolic actions
- organoleptic acceptability
- synthetic additives

REFERENCES

Ahmed, A.; Khan, K. A.; Ahmad, V. U.; Qazi, S. Antibacterial Activity of Juliflorine Isolated from *Prosopis juliflora*. *Planta Med.* **1986**, *4*, 285–288.

Ajila, C. M.; Aalami, M.; Leelavathi, K.; Rao, U. J. S. P. Mango Peel Powder: A Potential Source of Antioxidant and Dietary Fiber in Macaroni Preparations. *Innov. Food Sci. Emerg. Technol.* **2010**, *11*(1), 219–224.

Anila, L.; Vijayalakshmi, N. Flavonoids from *Emblica officinalis* and *Mangifera indica* Effectiveness for Dyslipidemia. *J. Ethnopharmacol.* **2002**, *79*(1), 81–87.

Apak, R.; Guclu, K.; Demirata, B.; Ozyurek, M.; Esin, C. S.; Bektasoglu, B.; Berker, K.; Ozyur, D. Comparative Evaluation of Various Total Antioxidant Capacity Assays

Applied to Phenolic Compounds with the CUPRAC Assay. *Molecules* **2007**, *12*, 1496–1547.

Arogba, S. S. Mango (*Mangifera indica*) Kernel: Chromatographic Analysis of the Tannin, and Stability Study of the Associated Polyphenol Oxidase Activity. *J. Food Compos. Anal.* **2000**, *13*, 149–156.

Ashour, M.; Wink, M. Biochemistry of Terpenoids: Monoterpenes, Sesquiterpenes and Diterpenes. In *Biochemistry of Plant Secondary Metabolism*; Wink, M. Ed.; John Wiley & Sons: New Jersey, 2010, Vol. 40.

Ayala-Zavala, J. F.; Vega-Vega, V.; Rosas-Domínguez, C.; Palafox-Carlos, H.; Villa-Rodriguez, J. A.; Siddiqui, M. W.; González-Aguilar, G. A. Agro-Industrial Potential of Exotic Fruit Byproducts as a Source of Food Additives. *Food Res. Int.* **2011**, *44*(7), 1866–1874.

Aydin, E.; Türkez, H.; Geyikoğlu, F. Antioxidative, Anticancer and Genotoxic Properties of α-Pinene on N2a Neuroblastoma Cells. *Biologia* **2013**, *68*, 1004–1009.

Babbar, N.; Oberoi, H. S.; Sandhu, S. K. Therapeutic and Nutraceutical Potential of Bioactive Compounds Extracted from Fruit Residues. *Crit. Rev. Food Sci. Nutr.* **2015**, *55*(3), 319–337.

Bedrossian, R. H. The Effect of Atropine on Myopia. *Ophthalmology* **1979**, *86*, 713–717.

Blakemore, P. R.; White, J. D. Morphine, the Proteus of Organic Molecules. *Chem. Commun.* **2002**, *11*, 1159–1168.

Birt, D. F.; Hendrich, S.; Wang, W. Dietary Agents in Cancer Prevention: Flavonoids and Isoflavonoids. *Pharmacol. Ther.* **2001**, *90*(2), 157–177.

Brinchmann, H. O.; Anmarkrud, N. Pilocarpine Medication in Open-Angle Glaucoma. A Study Using Pilocarpine Eyedrops and an Ocular Therapeutic System. *Acta Ophthalmol. (Copenh.)* **1979**, *57*, 55–62.

Burt, S. Essential Oils: Their Antibacterial Properties and Potential Applications in Foods—A Review. *Int. J. Food Microbiol.* **2004**, *94*, 223–253.

Chuarienthong, P.; Lourith, N.; Leelapornpisid, P. Clinical Efficacy Comparison of Anti Wrinkle Cosmetics Containing Herbal Flavonoids. *Int. J. Cosmet. Sci.* **2010**, *32*, 99–106.

Conte, A.; Scrocco, C.; Sinigaglia, M.; Del Nobile, M. A. Innovative Active Packaging Systems to Prolong the Shelf Life of Mozzarella Cheese. *J. Dairy Sci.* **2007**, *90*(5), 2126–2131.

Cordell, G. A. *The Alkaloids: Chemistry and Biology*. Elsevier: Amsterdam, 2001, Vol. 56, p 8.

Costa, B.; Trovato, A. E.; Comelli F.; Giagnoni, G.; Colleoni, M. The Non-Psychoactive Cannabis Constituent Cannabidiol Is an Orally Effective Therapeutic Agent in Rat Chronic Inflammatory and Neuropathic Pain. *Eur. J. Pharmacol.* **2007**, *556*, 75–83.

Cotelle, N. Role of Flavonoids in Oxidative Stress. *Curr. Topics Med. Chem.* **2001**, *1*, 569–590.

de Moraes Crizel, T.; Jablonski, A.; de Oliveira Rios, A.; Rech, R.; Flôres, S. H. Dietary Fiber from Orange Byproducts as a Potential Fat Replacer. *LWT—Food Sci. Technol.* **2013**, *53*(1), 9–14.

Croteau, R.; Kutchan, T. M.; Lewis, N. G. Natural Products (Secondary Metabolites). In *Biochemistry and Molecular Biology of Plants*; Buchanan, B., Gruissem, W., Jones, R. Eds.; American Society of Plant Physiologists: Rockville, MD, 2000, pp 1250–1318.

Cuthbert, W. Some Effects of Atropine on Smooth Muscle. *Br. J. Pharmacol. Chemother.* **1963**, *21*, 285–294.

Damianaki, A.; Bakogeorgou, E.; Kampa, M.; Notas, G.; Hatzoglou, A.; Panagiotou, S.; Gemetzi, C.; Kouroumalis, E.; Martin, P. M; Castanas, E. Potent Inhibitory Action of Red Wine Polyphenols on Human Breast Cancer Cells. *J. Cell. Biochem.* **2000**, *78*, 429–441.

Deka, B. C.; Sethi, V.; Parsad, R. et al. Application of Mixtures Methodology for Beverages from Mixed Fruit Juice/Pulp. *J. Food Sci. Technol.* **2001**, *38*(6), pp 615–618.

Dembitsky, V. M.; Poovarodom, S.; Leontowicz, H.; Leontowicz, M.; Vearasilp, S.; Trakhtenberg, S.; Gorinstein, S. The Multiple Nutrition Properties of Some Exotic Fruits: Biological Activity and Active Metabolites. *Food Res. Int.* **2011**, *44*(7), 1671–1701.

Dewick, P. M. *Medicinal Natural Products: A Biosynthetic Approach*, 2nd ed.; Wiley: Chicago, 2002.

Dorman, H.; Deans, S. Antimicrobial Agents from Plants: Antibacterial Activity of Plant Volatile Oils. *J. Appl. Microbiol.* **2000**, *88*(2), 308–316.

Eddleston, M.; Buckley, N. A.; Eyer, P.; Dawson, A. H. Management of Acute Organophosphorus Pesticide Poisoning. *Lancet* **2008**, *371*, 597–607.

Elleuch, M.; Bedigian, D.; Roiseux, O.; Besbes, S.; Blecker, C.; Attia, H. Dietary Fibre and Fibre-Rich By-products of Food Processing: Characterisation, Technological Functionality and Commercial Applications: A review. *Food Chem.* **2011**, *124*, 411–421.

Elson, C. E.; Maltzman, T. H.; Boston, J. L.; Tanner, M. A.; Gould, M. N. Anti-Carcinogenic Activity of D-limonene during the Initiation and Promotion/Progression Stages of DMBA-induced Rat Mammary Carcinogenesis. *Carcinogenesis* **1988**, *9*, 331–332.

Fujiki, H.; Horiuchi, T.; Yamashita, K.; Hakii, H.; Suganuma, M.; Nishino, H.; Sugimura, T. Inhibition of Tumor Promotion by Flavonoids. *Progress Clin. Biol. Res.* **1985**, *213*, 429–440.

Garattini, S.; Lamesta, L.; Mortari, A.; Valzelli, L. Pharmacological and Biochemical Effects of Some Reserpine Derivative. *J. Pharmacy Pharma* **1961**, *13*, 548–553.

Geller, R. J.; Lopez, G. P.; Cutler, S.; Lin, D.; Bachman, G. F.; Gorman, S. E. Atropine Availability as an Antidote for Nerve Agent Casualties: Validated Rapid Reformulation of High-Concentration Atropine from Bulk Powder. *Ann. Emerg. Med.* **2003**, *41*, 453–456.

Georgiev, V.; Ananga, A.; Tsolova, V. Recent Advances and Uses of Grape Flavonoids as Nutraceuticals. *Nutrients* **2014**, *6*(1), 391–415.

González-Montelongo, R.; Lobo, M. G.; González, M. Antioxidant Activity in Banana Peel Extracts: Testing Extraction Conditions and Related Bioactive Compounds. *Food Chem.* **2010**, *119*(3), 1030–1039.

Gorinstein, S.; Poovarodom, S.; Leontowicz, H.; Leontowicz, M.; Namiesnik, J.; Vearasilp, S.; Haruenkit, R.; Ruamsuke, P.; Katrich, E.; Tashma, Z. Antioxidant Properties and Bioactive Constituents of Some Rare Exotic Thai Fruits and Comparison with Conventional Fruits. In Vitro and In Vivo Studies. *Food Res. Int.* **2011**, *44*(7), 2222–2232 (this issue).

Gowri, B.; Platel, K.; Prakash, J. et al. Influence of Amla Fruits (*Emblica officinalis*) on the Bio-availability of Iron from Staple Cereals and Pulses. *Nutr. Res.* **2001**, *21*(12), 1483–1492.

Greenwood D. The Quinine Connection. *J. Antimicrob. Chemother.* **1992**, *30*, 417–427.

Greenspan, P.; Bauer, J. D.; Pollock, S. H.; Gangemi, J. D.; Mayer, E. P.; Ghaffar, A.; Hartle, D. K. Antiinflammatory Properties of the Muscadine Grape (*Vitis rotundifolia*). *J. Agric. Food Chem.* **2005**, *53*(22), 8481–8484.

Harborne, J. B.; Williams, C. A. Advances in Flavonoid Research Since 1992. *Phytochemistry* **2000**, *55*, 481–504.

Holdt S. L.; Kraan S. Bioactive Compounds in Seaweed: Functional Food Applications and Legislation. *J. Appl. Phycol.* **2011**, *23*, 543–597.

Holley, R. A.; Patel, D. Improvement in Shelf-Life and Safety of Perishable Foods by Plant Essential Oils and Smoke Antimicrobials. *Food Microbiol.* **2005**, *22*(4), 273–292.

Huang, M. T.; Ferraro, T. *Phenolic Compounds in Food and Cancer Prevention: Phenolic Compounds in Food and their Effects on Health II*; ACS Symposium Series 507, Chapter 2, 2009, pp 8–34.

Janevska, D. P.; Gospavic, R.; Pacholewicz, E.; Popov, V. Application of aHACCP-QMRA Approach for Managing the Impact of Climate Change on Food Quality and Safety. *Food Res. Int.* **2010**, *43*(7), 1915–1924.

Kale, A.; Gawande, S.; Kotwal, S. Cancer Phytotherapeutics: Role for Flavonoids at the Cellular Level. *Phytother. Res.* **2008**, *22*(5), 567–577.

Kanatt, S. R.; Chander, R.; Sharma, A. Antioxidant and Antimicrobial Activity of Pomegranate Peel Extract Improves the Shelf Life of Chicken Products. *Int. J. Food Sci. Technol.* **2010**, *45*(2), 216–222.

Kim, H. K.; Namgoong, S. Y.; Kim, H. P. Biological Actions of Flavonoids-I. *Arch. Pharmacol. Res.* **1993**, *16*, 18–27.

Kim, H. J.; Yokozawa, T.; Kim, H. Y. et al. Influence of Amla (*Emblica officinalis* Gaertn.) on Hypercholesterolemia and Lipid Peroxidation in Cholesterol-Fed Rats. *J. Nutr. Sci. Vitaminol.* **2005**, *51*(6), 413–418.

Knekt, P.; Kumpulainen, J.; Järvinen, R.; Rissanen, H.; Heliövaara, M.; Reunanen, A.; Aromaa, A. Flavonoid Intake and Risk of Chronic Diseases. *Am. J. Clin. Nutr.* **2002**, *76*(3), 560–568.

Knoblocha, K.; Paulia, A.; Iberla, B.; Weiganda, H.; Weisa, N. Antibacterial and Antifungal Properties of Essential Oil Components. *J. Essent. Oil Res.* **1989**, *1*, 119–128.

Kole P. L.; Jadhav H. R.; Thakurdesai P.; Naik, N. A. Cosmetics Potential of Herbal Extracts. Nat. Prod. Radian. **2005**, *4*(4), 315–321.

Kris-Etherton P. M.; Lefevre M.; Beecher G. R.; Gross M. D.; Keen C. L.; Etherton T. D. Bioactive Compounds in Nutrition and Health—Research Methodologies for Establishing Biological Function: The Antioxidant and Anti-Inflammatory Effects of Flavonoids on Atherosclerosis. *Annu. Rev. Nutr.* **2004**, *24*, 511–538.

Larrauri, J. A. New Approaches in the Preparation of High Dietary Fibre Powders from Fruit by-Products. *Trends Food Sci. Technol.* **1999**, *10*(1), 3–8.

Lorenzetti, B. B.; Souza, G. E.; Sarti, S. J.; Santos, F. D.; Ferreira, S. H. Myrcene Mimics the Peripheral Analgesic Activity of Lemongrass Tea. *J. Ethnopharmacol.* **1991**, *34*(1), 43–48.

Lucas, D. M.; Still, P. C.; Perez, L. B.; Grever, M. R.; Kinghorn, A. D. Potential of Plant-Derived Natural Products in the Treatment of Leukemia and Lymphoma. *Curr. Drug Targets* **2010**, *11*, 812–822.

Mack, A. H.; Harrington, A. L.; Frances, R. J. *Clinical Manual for Treatment of Alcoholism and Addictions,* 1st ed. American Psychiatric Publishing, Inc., 2010, p 304.

Mahato, S. B.; Sen, S. Advances in Triterpenoid Research, 1990–1994. *Phytochemistry* **1997**, *44*, 1185–1236.

Manfred, H. Alkaloids: Nature's Curse or Blessing?. Wiley-VCH, 2002.

Marín, F. R.; Soler-Rivas, C.; Benavente-García, O.; Castillo, J.; Pérez-Alvarez, J. A. By-products from Different Citrus Processes as a Source of Customized Functional Fibres. *Food Chem.* **2007**, *100*, 736–741.

Mukherjee P. K.; Maity, N.; Nemaa N. K.; Sarkar B. K. Bioactive Compounds from Natural Resources against Skin Aging. *Phytomedicine* **2011**, *19*, 64–73.

Narayana, K. R.; Reddy, M. S.; Chaluvadi, M. R.; Krishna, D. R. Bioflavonoids Classification, Pharmacological, Biochemical Effects and Therapeutic Potential. *Indian J. Pharmacol.* **2001**, *33*, 2–16.

O'Shea, N.; Arendt, E. K.; Gallagher, E. Dietary Fibre and Phytochemical Characteristics of Fruit and Vegetable By-Products and Their Recent Applications as Novel Ingredients in Food Products. *Innov. Food Sci. Emerg. Technol.* **2012**, *16*, 1–10.

Palafox-Carlos, H.; Ayala-Zavala, F.; González-Aguilar, G. A. The Role of Dietary Fiber in the Bioaccessibility and Bioavailability of Fruit and Vegetable Antioxidants. *J. Food Sci.* **2010**, *76*(1), R6–R15.

Parvez S.; Kang M.; Chung H.-S.; Bae H. Naturally Occurring Tyrosinase Inhibitors: Mechanism and Applications in Skin Health, Cosmetics and Agriculture Industries. *Phytother. Res.* **2007**, *21*, 805–816.

Patel, S.; Goyal, G. *Emblica officinalis* Geart: A Comprehensive Review on Phytochemistry, Pharmacology and Ethnomedicinal Uses. *Res. J. Med. Plant* **2012**, *6*(7), 6-16.

Peñuelas, J.; Munné-Bosch, S. Isoprenoids: An Evolutionary Pool for Photoprotection. *Trends Plant Sci.* **2005**, *10*, 166–169.

PhytoCellTec™. *Solar Vitis/Vitis vinifera*. Available online: www.phytocelltec.ch/pctvitis-vinifera.php (accessed on October 01, 2013).

Puranik, V; Srivastava, P; Mishra, V. et al. Effect of Different Drying Techniques on the Quality of Garlic: A Comparative Study. *Am. J. Food Technol.* **2012**, *7*(50), 311–319.

Rabi, T.; Bishayee, A. Terpenoids and Breast Cancer Chemoprevention. *Breast Cancer Res Treat.* **2009**, *115*, 223–239.

Rahman, A; Mohammad, I. C. *Frontiers in Drug Design and Discovery*, Volume 6, 2014.

Rea, G.; Antonacci, A.; Lambreva, M.; Margonelli, A.; Ambrosi, C.; Giardi, M. T. The NUTRASNACKS Project: Basic Research and Biotechnological Programs on Nutraceutical. In *Bio-Farms for Nutraceuticals: Functional Food and Safety Control by Biosensors*; Giardi, M. T., Rea, G., Berra, B., Eds.; Springer: US, 2010, Vol. 698, Chapter 1, pp 1–16.

Runowicz, C. D.; Wiernik, P. H.; Einzig, A. I.; Goldberg, G. L.; Horwitz, S. B. Taxol in Ovarian Cancer. *Cancer* **1993**, *71*, 1591–1596.

Sacchettini, J. C.; Poulter, C. D. Creating Isoprenoid Diversity. *Science* **1997**, *277*, 1788–1789.

Sawant L. P. Quantitative HPLC Analysis of Ascorbic Acid and Gallic Acid in *Phyllanthus emblica*. *J. Anal. Bioanal. Technol.* **2010**.

Scartezzini, P.; Speroni, E. Review on Some Plants of Indian Traditional Medicine with Antioxidant Activity. *J. Ethnopharmacol.* **2000**, *71*(1), 23–43.

Schieber, A.; Stintzing, F. C.; Carle, R. By-Products of Plant Food Processing as a Source of Functional Compounds—Recent Developments. *Trends Food Sci. Technol.* **2001,** *12*(11), 401–413.

Schmidt B. M.; Ribnicky D. M.; Lipsky P. E.; Raskin I. Revisiting the Ancient Concept of Botanical Therapeutics. *Nat. Chem. Biol.* **2007,** *3*, 360–366.

Schwaab, B.; Katalinic,A.; Böge, U. M.; Loh, J.; Blank, P.; Kölzow, T.; Poppe, D.; Bonnemeier, H. Quinidine for Pharmacological Cardioversion of Atrial Fibrillation: A Retrospective Analysis in 501 Consecutive Patients. *Ann. Noninvasive Electrocardiol.* **2009,** *14*, 128–136.

Shan, B.; Cai, Y. Z.; Brooks, J. D.; Corke, H. The In Vitro Antibacterial Activity of Dietary Spice and Medicinal Herb Extracts. *Int. J. Food Microbiol.* **2007,** *117*(1), 112–119.

Siliha, H.; El-Sahy, K.; Sulieman, A.; Carle, R.; El-Badawy, A. Citrus Wastes: Composition, Functional Properties and Utilization. *Obst-, Gemüse- und Kartoffelverarbeitung [Fruit, Veg. Potato Process.]* **2000,** *85*, 31–36.

Stintzing, F. C.; Carle, R. Functional Properties of Anthocyanins and Betalainsinplants, Food, and in Human Nutrition. *Trends Food Sci. Technol.* **2004,** *15*(1), 19–38.

Sun, J. D-Limonene: Safety and Clinical Applications. *Altern. Med. Rev.* **2007,** *12*, 259–264.

Ultee, A.; Bennik, M. H. J.; Moezelaar, R. The Phenolic Hydroxyl Group of Carvacrol is Essential for Action against the Food-Borne Pathogen Bacillus cereus. *Appl. Environ. Microbiol.* **2002,** *68*(4), 1561–1568.

Van Dam, R. M.; Willett, W. C.; Manson, J. E.; Hu, F. B. Coffee, Caffeine, and Risk of Type 2 Diabetes: A Prospective Cohort Study in Younger and Middle-Aged U.S. Women. *Diabetes Care* **2006,** *29*, 398–403.

Veselovskaya, N. V.; Kovalenko, A. E. *Drugs: The Properties, Action, Pharmacokinetic, Metabolism.* Triada-X: Moscow, 2000a, p 75.

Veselovskaya, N. V.; Kovalenko, A. E. *Drugs: The Properties, Action, Pharmacokinetic, Metabolism.* Triada-X: Moscow, 2000b, p 136.

Veselovskaya, N. B.; Kovalenko, A. E. Effects of Alkaloids. *Drugs* **2000c,** *9*, 11–12.

Veselovskaya, N. V.; Kovalenko, A. E. *Drugs: The Properties, Action, Pharmacokinetic, Metabolism.* Triada-X: Moscow, 2000d, pp 51–52.

Wattenberg, L. W. Effects of Dietary Constituents on the Metabolism of Chemical carcinogens. *Cancer Res.* **1975,** *35*, 3326–3330.

Withers, S. T.; Keasling, J. D. Biosynthesis and Engineering of Isoprenoid Small Molecules. *Appl. Microbiol. Biotechnol.* **2007,** *73*, 980–990.

Wu, C. Q.; Chen, F.; Wang, X.; Kim, H. J.; He, G. Q.; Haley-Zitlin, V.; Huang, G. Antioxidant Constituents in Feverfew (*Tanacetum parthenium*) Extract and their Chromatographic Quantification. *Food Chem.* **2006,** *96*, 220–227.

Zhang, Y.; Seeram, N. P.; Lee, R.; Feng, L.; Hebe, D. Isolation and Identification of Strawberry Phenolics with Antioxidant and Human Cancer Cell Antiproliferative Properties. *J. Agric. Food Chem.* 2008, *56*, 670–675.

INDEX

Printed and bound by CPI Group (UK) Ltd, Croydon, CR0 4YY

23/10/2024

01777704-0002